Encyclopaedia of
Mathematical Physics

Encyclopaedia of Mathematical Physics

Volume 1

Mathematical Physics

Manpal Singh

ANMOL PUBLICATIONS PVT. LTD.
NEW DELHI - 110 002 (INDIA)

ANMOL PUBLICATIONS PVT. LTD.
4374/4B, Ansari Road, Daryaganj
New Delhi - 110 002
Ph.: 23261597, 23278000
Visit us at: www.anmolpublications.com

Encyclopaedia of Mathematical Physics

First Published, 2006

ISBN 81-261-2611-6 (Set)

PRINTED IN INDIA

Published by J.L. Kumar for Anmol Publications Pvt. Ltd., New Delhi - 110 002 and Printed at Mehra Offset Press, Delhi.

Contents

Preface

Science is the sermon of today. In every walk of life, since childhood till old age, science plays an important role in the life of man. Uptill some decades back, studied merely as a subject in schools and colleges, Science has assumed the status of an essential part of life today. With the expansion of literacy among the masses and growing awareness, they have woken up from the slumber of ignorance and have started taking up things more sincerely and thoughtfully. It is one of the reasons that our nation is heading towards new goals day after day.

It is a vast subject to study and talk about. For the purpose of ease, Science has been divided into more than two hundred and fifty branches. Specialisation in any branch can be obtained at different institutions and universities, scattered all over the country and abroad.

Apart from other super-specific branches, Physics, Chemistry, Zoology, Botany, Mathematics and Computronics, etc. are more general and closer to daily human life. From the point of view of the cut-throat competition in educational world, a thorough study of the above-mentioned branches, proves very handy, for the new aspirants, especially and for teachers and tutors, generally.

An urgent need was being felt for such a series of books, encompassing the commonly-studied and discussed branches of Science of equal significance and presented in an easy language, understood by all. With God's grace, our dream came true and we are happy, while introducing the dream-series for the lovers of knowledge.

The series consists of many a title, covering various branches of Science, such as Physics, Chemistry, Zoology, Botany, Geology, Agriculture Science, Computer Science, Mathematics and so on. The series is important and a must not only for students but also for teachers. Explained in simple language and aided with graphics and diagrams, everything, between the two ends—title and back cover—of the series, has been presented in a perfect way.

This set of books is devoted to **Mathematical Physics**, which comprises two volumes: (1) Mathematical Physics and (2) Mathematical Coordinations and Variations.

It is hoped that the series would serve both, the students community and the academics, equally and will prove to be a boon for those, aspiring for newer heights. Nevertheless, any suggestions for further betterment of the endeavour will be highly appreciated and solicited.

Editor

1

Position Coordination

Phase Space Representation

In classical mechanics, the instantaneous dynamical state of a particle is completely specified by its three position coordinates x, y, z, and the corresponding momentum components p_x, p_y, p_z. Thus, six coordinates are needed to specify a one-particle system completely.

Gibbs suggested that any instantaneous state (position and momentum) of the particle may be conveniently represented by some point in an imaginary six-dimensional space in which the six coordinates x, y , z , p_x, p_y, p_z are marked along six mutually perpendicular axes in space. This six-dimensional space is known as 'phase space' or 'μ-space'. The point in the phase space representing the instantaneous state of the particle is called the 'phase point'. As the time progresses, the phase point moves in the phase space. The path of the phase point represents the 'trajectory' of the particle.

If the system contains a large number of particles, then every particle is completely specified by a point in the phase space. Thus, the instantaneous state of a system of particles is represented by the corresponding distribution of phase points in the phase space.

Division of Phase Space into Cells: The conception of a "point" in the phase space (position-momentum space) is to be considered in the light of uncertainty principle. The phase space is divided into tiny six-dimensional cells whose sides are δx, δy, δz, δp_x, δp_y, δp_z. Such cells are called 'phase cells'. The volume of each of these cells is

$$\delta\tau = \delta x\,\delta y\,\delta z\,\delta p_x\,\delta p_y\,\delta p_z .$$

Classically there is no restriction on the volume of the phase cell; it may be reduced to any extent, tending to zero also, without affecting the classical results. However, according to uncertainty principle

$$\delta x\,\delta p_x \geq h/2\pi$$
$$\delta y\,\delta p_y \geq h/2\pi$$
$$\delta z\,\delta p_z \geq h/2\pi .$$

Hence we see that

$$\delta\tau \geq (h/2\pi)^3$$

Thus, a "point" in the phase space is actually a cell whose minimum volume is of the order of h^3. It means that a particle in phase space cannot be considered exactly located at the point x , y , z , p_x , p_y, p_z; but can only be found somewhere within a phase cell centered at that point.

Now, the state of a system can be described by specifying the distribution of the particles of the system among the phase cells. We can determine the probabilities of occurrence of all possible distributions that are permitted by the nature of the system. Out of these we can select the most probable one. *The state of the system when it is in thermal equilibrium corresponds to the most probable distribution of particles in the phase space.*

Microstates and Macrostates of a System : Let us consider a system consisting of a large number of gas molecules in the phase space which has been divided into tiny cells. Each cell represents a small region of position and momentum. Each molecule may be specified by a point (phase point) lying somewhere inside one of these cells.

The microstate of the system at a particular instant can be defined when we specify as to which particular cell each molecule of the system belongs at that instant. This deep information is, however, unnecessary to determine the observable properties of the system (gas). For example, the density is uniform if the number of molecules in each cell is same, regardless of which particular molecule lies in which particular cell.

A macrostate of the system, on the other hand, can be defined by just giving the number of molecules in each cell; such as n_1 molecules are in cell 1, n_2 are in cell 2, and so on.

There may be a large number of microstates corresponding to the same macrostate. As an example, let us consider a system of three molecules only (for simplicity) named *a, b, c*; which are to be distributed in two halves of a box, the left half L and the right half *R*. There are four possible distributions:

(i) 3 molecules in *L* and 0 in *R* ,

(ii) 2 molecules in *L* and 1 in *R* ,

(iii) 1 molecule in *L* and 2 in *R* ,

(iv) 0 molecule in *L* and 3 in *R* .

Let us call these distributions as (3,0); (2, 1); (1, 2); (0, 3) respectively.

Now, the distributions (3, 0) and (0, 3) can occur in one way only; while the distributions (2, 1) and (1, 2) can each occur in three ways as illustrated in the following table :

Distribution	Left (L)	Right (R)
(3, 0)	*abc*	—
	ab	*c*
(2, 1)	*ac*	*b*
	bc	*a*
	a	*bc*
(1, 2)	*b*	*ac*
	c	*ab*
(0, 3)	—	*abc*

Thus, the total number of ways in which three molecules can occupy two halves of the box are 1 + 3 + 3+1 = 8 (= 2^3) corresponding to four different distributions. Each way of arrangement of molecules is a microstate of the system, while each different distribution of molecules is a macrostate. Thus, there are eight microstates and four macrostates of the system. There is only one microstate *(abc,—)* corresponding to the macrostate (3, 0); three microstates *(ab, c)*; *(ac, b)*; *(bc, a)* corresponding to the macrostate (2,1); and so on.

In a gaseous system there is a very large number of molecules which are in random motion. Hence there exists a very large number of microstates corresponding to a given macrostate. If at two instants t and t' (say) the experimentally measurable quantities pressure, volume and temperature are same, then the system is in the same macroscopic state at these instants. However, the microscopic states may be (and are) different because the positions and velocities (momenta) of the individual molecules are constantly changing.

Constraints and Accessible States

The restrictions imposed by physical laws on the distribution of molecules among the cells in the phase space are called the 'constraints' of the system. For example, the total number of molecules N (say) in a system remains unchanged. If there are n_1 molecules in cell 1, n_2 in cell 2, ..., and n_r in cell r, then the constraint requires

$$\sum_{i=1}^{r} n_i = N.$$

Similarly, if the total energy E of the system is constant and if n_1 molecules have energy ε_1 each, n_2 molecules have energy ε_2 each, ..., then we have the constraint

$$\sum_{i=1}^{r} n_i \varepsilon_i = E.$$

The microstates which are permitted under the constraints imposed upon the system are called 'accessible microstates'.

A fundamental postulate of statistical mechanics is that *for a system in equilibrium, all accessible microstates corresponding to a given macrostate are equally probable.* This is known as the postulate of 'equal a priori probability'.

Phase-space Diagram of one-dimensional Oscillator: An oscillator is a particle bound to a centre by a harmonic force, and so it has kinetic as well as potential energy.

The energy of one-dimensional oscillator is

$$E = \frac{P_x^2}{2m} + \frac{1}{2}kx^2,$$

where m is the mass of the oscillator and k is the spring factor of the restoring force acting on the oscillator. The given equation can be written as

$$\frac{Px^2}{2mE} + \frac{kx^2}{2E} = 1$$

or
$$\frac{x^2}{(2E/k)} + \frac{P_x^{\,2}}{2mE} = 1.$$

In the x-p_x phase space, this represents an ellipse whose semi-major axis $a = \sqrt{2E/k}$ and semi-minor axis $b = \sqrt{2mE}$.

The instantaneous state (position-momentum) of the oscillator is represented by some point on the ellipse. Thus, the trajectory of the oscillator in the two-dimensional phase space is ellipse.

The area of the ellipse in the phase space is given by

$$A = \pi ab = \pi \sqrt{2E/k}\,\sqrt{2mE} = 2\pi E\sqrt{m/k}.$$

This area represents the classical phase space available to the oscillator having energy between 0 and E. Hence the phase space available to the oscillator having energy between E and $E + \delta E$ will be equal to the differential of the above area, that is,

$$\delta A = 2\pi\sqrt{m/k}\,\delta E.$$

This represents the region of states accessible to the oscillator.

Phase-space Diagram for a Single Particle free to Move in One-dimension : Let us consider a single particle moving freely (under no forces) in one dimension within a box of length L. Suppose the box lies between $x = 0$ and $x = L$, and the particle moves along the x-axis. Then, the instantaneous state of the particle is specified by a position coordinate x and a momentum coordinate p_x. Thus, the phase space is just the two-dimensional $(x — p_x)$ space.

Since the particle is assumed to be "free", its energy E is merely its kinetic energy so that

$$E = \frac{1}{2}mv_x^2 = \frac{px^2}{2m} \qquad [\because P_x = mv_x]$$

or

$$p_x = \pm\sqrt{2mE} \qquad \text{...(i)}$$

If the energy of the particle has to be between E and $E + \delta E$, then its momentum must lie in some small range δp_x about the possible values $p_x = \pm\sqrt{2mE}$. The region of phase space accessible to the particle is then the one shown by shaded areas.

If the (two-dimensional) phase space be divided into tiny cells, each of area

$$\delta x \;\; \delta p_x = h \,,$$

then this region would contain a large number of cells. These cells would represent the accessible states in which the particle can be found.

The total area of the accessible region (shaded) is $2L \;\delta p_x$. The differential δp_x can be computed from eq. (i):

$$\delta p_x = \frac{1}{2}(2mE)^{-1/2} 2m\delta E = \frac{1}{2}(2m)^{1/2} E^{-1/2} \delta E$$

$$\therefore \text{ total area} = 2L\delta p_x = L(2m)^{1/2} E^{-1/2} \delta E.$$

Since the area of a single phase cell is h, the number of phase cells for the particle in the energy range E to $E + \delta E$ is

$$\frac{L}{h}(2m)^{1/2} E^{-1/2} \delta E.$$

The particle is equally likely to be found in any of the equal-sized cells. Therefore, the number of microstates $\Omega (E)$ of the particle is equal to the number of phase cells in the allowed region. That is,

$$\Omega(E) = \frac{L}{h}(2m)^{1/2} E^{-1/2} \delta E.$$

Number of Phase Cells in a given Energy Range for a Three-dimensional Free Particle : For a single particle free to move in a three-dimensional space we have a six-dimensional phase space. The microstate of the particle is specified by three position coordinates x , y , z and three momentum coordinates p_x , p_y, p_z .

The volume of a phase cell in the phase space is

$$= \delta x \, \delta y \, \delta z \;\, \delta p_x \, \delta p_y \, \delta p_z$$

$$= (\delta x \, \delta p_x)(\delta y \, \delta p_y)(\delta z \, \delta p_z)$$

$$= h^3 .$$

The total volume of the phase space is

$$\iiiint\!\!\iint dx\, dy\, dz\, dp_x\, dp_y\, dp_z.$$

We have $\iiint dx\, dy\, dz$ = given volume V.

$\therefore$ volume of the phase space $= V \iiint dp_x\, dp_y\, dp_z.$ ...(i)

Let, in the energy range 0 to E, the possible values of momentum be from 0 to p. Then, we can write for the volume of the momentum space, which is a sphere of radius p,

$$\iiint dp_x,\ dp_y,\ dp_z = \frac{4}{3}\pi (p_{\max})^3 \qquad \text{...(ii)}$$

The energy E of a 'free' particle is merely its kinetic energy so that

$$E = \frac{p^2}{2m} \qquad [E = \frac{1}{2}mv^2, p = mv]$$

or

$$(p_{\max})^2 = (px^2 + py^2 + pz^2) = 2mE. \qquad \text{...(iii)}$$

Using eq. (ii) and (iii), we can write eq. (i) as :

$$\text{volume of the phase space} = V\left(\frac{4}{3}\pi\right)(2mE)^{3/2}.$$

Since the volume of a phase cell is h^3, the number of cells (in energy range 0 to E) in the phase space is

$$\frac{4\pi V}{3h^3}(2mE)^{3/2}$$

This is the required result.

For a single particle, the number of accessible microstates is equal to the number of cells in the phase space. Therefore, the number of microstates in the energy range 0 to E is given by

$$\phi(E) = \frac{4\pi V}{3h^3}(2mE)^{3/2}$$

The number of microstates (i.e. number of phase cells) in

the energy range E to $E + \delta E$ is obtained by differentiating the above equation with respect to E. Thus

$$\Omega(E) = \left[\frac{\partial\phi(E)}{\partial E}\right]\delta E = \frac{4\pi V}{3h^3}\frac{3}{2}(2mE)^{1/2}2m\,\delta E$$

or
$$\Omega(E) = \frac{4\sqrt{2}\,\pi V}{h^3}m^{3/2}E^{1/2}\,\delta E.$$

This is the expression for a single particle in a three-dimensional box.

Number of Phase Cells in a given Energy range for a Harmonic Oscillator : For a one-dimensional oscillator, the phase space is two-dimensional (x — p_x). The area of a phase cell in this phase space is

$$\delta x\,\delta p_x = h\,. \qquad \text{(by uncertainty principle)}$$

The (total) energy of the oscillator of mass m is given by

$$E = \frac{p_x^2}{2m} + \frac{1}{2}kx^2$$

which may be written as

$$\frac{x^2}{(2E/k)} + \frac{p_x^2}{2mE} = 1.$$

This represents an ellipse of semi-major axis $a = \sqrt{2E/k}$ and semi-minor axis $b = \sqrt{2mE}$. Thus, the area of the phase space of the oscillator having energy between 0 and E is

$$\pi ab = \pi\sqrt{2E/k}\,\sqrt{2mE} = 2\pi E\sqrt{m/k}\,.$$

The frequency of the oscillator is ν, and so

$$\nu = \frac{1}{2\pi}\sqrt{\frac{k}{m}}.$$

$\therefore$ area of phase space $= 2\pi E\dfrac{1}{2\pi\nu} = E/\nu\,.$

Since the area of each phase cell is h, the number of phase cells (in energy range 0 to E) is

$$\frac{E/\nu}{h} = \frac{E}{h\nu}$$

Number of (Micro) States Accessible to a Macroscopic System: Let us consider a macroscopic system whose external parameters are given so that its energy states are determined. Let E be the total energy of this system. Let us subdivide the energy scale of the system into equal, small intervals of fixed magnitude δE. The energy interval is so chosen that it is very small compared to the total energy of the system, but quite large compared to the energy of a single particle in the system and thus quite large compared to the separation in energy between adjacent energy states of the system. Any interval δE thus contains many possible quantum states of the system.

Let $\Omega(E)$ be the number of states with energies lying in the interval between E and $E + \delta E$. This number $\Omega(E)$ depends on the magnitude δE chosen. Since $\delta E << E$, $\Omega(E)$ must be simply proportional to δE. So we can write

$$\Omega(E) = \rho(E)\delta E.$$

The quantity $\rho(E)$ is called the 'density of states' because it is equal to the number of states per unit energy range at the given energy E. Further, $\rho(E)$ is independent of the size of δE.

Now, it is possible to obtain $\Omega(E)$ if we know the quantity $\Phi(E)$, the total number of states having energies *less* than E.

The number $\Omega(E)$ of states having energies lying between E and $E+\delta E$ is then simply given by

$$\Omega(E) = \Phi(E+\delta E) - \Phi(E) = \frac{d\Phi}{dE}\delta E.$$

This is the required expression.

Number of States of a Single Particle in a One-dimensional Box: Let there be a single particle of mass m free to move in a one-dimensional box of length L. The possible energy values of this system are given by

$$E = \frac{h^2}{8mL^2} n^2; \qquad n = 1, 2, 3, 4, \ldots$$

The coefficient $h^2/8mL^2$ is very small if L is of directly measurable size ($\approx 10^{-25}$ joule for a nitrogen molecule moving in a box of length L = 1 cm). The quantum number n is thus very large ($\approx 10^9$, in case of nitrogen molecule whose average energy at room temperature is of the order of 10^{-7} joule).

From the last expression, the value of n for a given energy E is

$$n = \frac{2L}{h}(2mE)^{1/2}.$$

The successive quantum states correspond to values of n differing by unity. Therefore, the total number $\Phi(E)$ of quantum states having energies less than E, or quantum numbers less than n, is simply $(n-1)$ or n (because n is very large). Thus

$$\Phi(E) = n = \frac{2L}{h}(2mE)^{1/2}.$$

Now, the number of states having energies between E and $E + \delta E$ is given by

$$\Omega(E) = \frac{d\Phi}{dE}\delta E$$

$$= \frac{d}{dE}\left\{\frac{2L}{h}(2mE)^{1/2}\right\}\delta E$$

$$= \left\{\frac{2L}{h}\frac{1}{2}(2mE)^{-1/2}2m\right\}\delta E$$

or

$$\Omega(E) = \frac{L}{h}(2m)^{1/2}E^{-1/2}\delta E$$

This is the number of states accessible to a free particle confined in a box of length L.

PROBLEM

40 identical marbles are thrown into a box with 4 similar compartments, (i) Calculate the number of microstates in each of the macrostates (11, 10, 9, 10); (1, 1, 1, 37); (0, 0, 0, 40) and (10, 10, 10, 10). (ii) State the macrostate for which the probability of occurrence is maximum.

Solution :

(i) If, in a particular macrostate of a system of N particles, there are n_1, n_2, ..., n_r particles in the first, second, ...rth cell respectively, then the number of microstates corresponding to this macrostate is

$$\frac{N!}{n_1!\ n_2!\ \ldots\ldots\ n_r!}.$$

Macrostate	No. of microstates
(11,10, 9, 10)	$\frac{40!}{11!\ 10!\ 9!\ 10!}$
(1,1, 1, 37)	$\frac{40!}{1!\ 1!\ 1!\ 37!} = 40 \times 39 \times 38 = 59280.$
(0,0, 0, 40)	$\frac{40!}{0!\ 0!\ 0!\ 40!} = 1$
(10,10, 10, 10)	$\frac{40!}{10!\ 10!\ 10!\ 10!}$

(ii) Since the compartments are similar (no one is favoured), the maximum probability is for equal distribution of marbles in them. Thus, the maximum probable macrostate is (10, 10, 10, 10).

2

Natural Process and Disorder

We know from experience that it is easy to pass from an ordered state into a disordered state, but the reverse process is much more difficult. Suppose we have a box divided into two compartments which are open to each other, and we put a number of red balls in one compartment and a number of black balls in the other. This in an ordered arrangement of balls. If we now shake the box a number of times, many red balls will go into the second compartment and many black balls into the first. Thus, we easily pass from an ordered state to a disordered state. But now if we shake the box in an attempt to bring once again all the red balls in one compartment and all the black ones in other, we shall, practically cent percent fail.

That is, it is extremely difficult , if not impossible, to get the balls once more in an ordered state. Thus, common experience tells us that the disordered states are the most probable in nature while the ordered states are the least probable. In fact, *in all natural processes there is a tendency to proceed towards a state of greater disorder*. We justify this statement by the following examples:

(i) *Free Expansion:* Suppose a gas is filled in a vessel *A* which is connected to another 'evacuated' vessel *B* through a stop-cock. When the stop-cock is suddenly opened, the gas rushes into the vacuum of *B*. Before the expansion, the molecules were occupying the vessel *A*, but after the expansion they have spread over the vessel *B* also. In a sense they have become more disorderly.

The point becomes more clear in the mixing of gases. Suppose, the vessel *A* contained oxygen and the vessel *B* contained nitrogen. This is an ordered arrangement in the sense that the gas molecules are sorted according to their type. On opening the stop-cock, the gases become mixed because of the kinetic motion of the molecules. Complete disorder results; there is now no way to distinguish between one gas and the other.

(ii) *Heat Conduction :* Suppose two bodies at temperatures T_1 and T_2 are put in contact so that, due to heat-conduction, they attain a common temperature T_m. Before conduction, we could classify between the molecules of body *A* and those of body *B* because the former were at temperature T_1 and the latter at T_2 *i.e.* they had different *rms* speeds. But, after conduction, this classification is lost because all the molecules of the system are at the same temperature T_m and have the same *rms* speed. Thus, the system has passed to a greater disorderly state.

Entropy and Disorder

We find a beautiful analogy between 'entropy' and 'degree of disorder' of a system. Just as the entropy of a system increases due to a natural process taking place in the system, the degree of molecular disorder of the system also increases. In other words, we may state that the *entropy of a system is a measure of the degree of molecular disorder existing in the system.* This is Boltzmann's statistical interpretation of entropy. Boltzmann has given a mathematical equation relating entropy and disorder :

$$\boxed{S = k \log_e w,}$$

where k is a constant and w is the disorder parameter.

Application : The relation between the entropy and the molecular disorder of a system has been applied in the adiabatic demagnetisation method of obtaining very low temperatures.

The atoms of a paramagnetic salt are small magnets; and normally they are in a disordered state. When the salt is placed in a magnetic field, all the atoms are aligned in an ordered state parallel to the lines of force. Thus, the entropy due to their positional disorder decreases. But at the same time, the entropy due to the thermal motion of the atoms must increase (because there can be no net decrease in the entropy of the system) so that heat is developed. This heat is removed out by placing the salt in contact with liquid-helium and the salt is then insulated from the surroundings. Now, when the magnetic field is withdrawn, the positional entropy increases. This will be so at the cost of entropy due to the thermal motion which thus decreases and the salt is cooled.

Heat : Disordered Energy

By disorder, we mean randomness, or chaos, or molecular disorganisation. We know that heat in a body is the kinetic energy associated with random motions of the molecules constituting the body. Heating a body means increasing the molecular agitation in the body *i.e.* increasing the molecular disorder in the body. Thus, heat is the energy of disordered molecules, or heat is disordered energy.

Infact, the conversion of mechanical energy into heat energy is the conversion of a state of orderly motion into a state of disorderly motion. For example, when a bullet fired from a gun strikes a target, both the bullet and the target become hot. That is, the orderly motion of the bullet is converted into disorderly motion of the molecules of the bullet and of the target.

'Heat' and 'work' (mechanical energy) are two *different* forms of energy. The basic difference between them is that 'heat is disordered energy', whereas 'work is ordered energy'. For example, the heat energy of a gas is the energy of disorderly motion of its molecules. As such, the gas cannot do work. But when the gas molecules , or majority of them , are made to move in an orderly manner in a certain direction , the gas can do work, e.g. it can drive a piston in a cylinder. Thus , work is associated with a well-organised motion of the body as a whole. Now, the job of a heat engine is to convert heat into work, and the conversion of heat into work is, infact, the conversion of a disordered mechanical motion into organised mechanical motion.

Melting of ice results in the freedom of molecules to move about. Thus, it corresponds to an increase in disorder.

When work is done against the friction between two sliding surfaces, it is converted into heat *i.e.* the sliding surfaces become hot. That is, the work is converted into the random motion of the molecules in the surfaces. The increase in the molecular disorder is always associated with an increase in the entropy.

The natural tendency toward an increase in entropy, and thus in disorder, is a matter of everyday observation. Dry leaves turn to dust, a book turns into a mass of pulp when left out in the rain. Some processes in plants and animals, however, seem to violate the second law of thermodynamics since they start with unstructured materials and create highly organised configurations. This is an increase in the order of the system, corresponding to a decrease in entropy. But these are not *isolated* systems. A detailed analysis shows that considering the system plus the surroundings, there is always a net loss of order (or net increase in entropy). A green plant, for example, needs sunlight to live. A decrease in its entropy is always accompanied by a *greater* increase in the entropy of the sun. When we put bricks together to build a house , there is a decrease in entropy, but at the same time we have done work which has been converted into heat. That is, there is an increase in the entropy of the agent putting the bricks together. In general, whenever there is a decrease in the entropy of a system, we are sure that there is greater gain also somewhere else in the universe. Thus, such processes are consistent with the second law of thermodynamics.

3

Radiant Energy

Radiation : The process by which heat is transferred directly from one body to another without affecting the intervening medium, is called "radiation." It is by radiation that the heat from the sun reaches the earth.

To explain heat-transfer by radiation, it is universally assumed that all bodies at all times are emitting energy. This energy is called "radiant energy" or "thermal radiation" and is in the form of electromagnetic waves. These waves travel with the velocity of light and are transmitted through vacuum or through a medium like air. When they fall on a body which is not transparent to them, they are absorbed and their energy is converted into heat.

The thermal radiation emitted by a body, per unit time and per unit surface area, depends on the nature of its surface and on its temperature. At low temperatures the rate of emission is small. As the temperature is increased, the rate of emission increases rapidly in proportion to the fourth power of the absolute temperature. Further, the thermal radiation emitted by a body is a mixture of waves of different wave-lengths. At ordinary and moderately high temperatures mostly the longer waves (infra-red) are emitted, but at very high temperatures shorter waves are also emitted.

Properties of Thermal Radiation : The thermal radiation is completely identical to light.

(i) It travels through empty space with the same velocity as light.

(ii) It travels in straight lines in the same sense as light.

(iii) It obeys the inverse-square law.

(iv) It undergoes reflection, refraction and total internal reflection obeying the same laws as light.

(v) It exhibits interference, diffraction and polarisation.

(vi) It exerts a small, but finite, pressure on the surface on which it is incident.

Prevost's Theory of Exchanges : Prevost, in 1792, enunciated a "theory of exchanges" to account for the phenomenon of radiation. According to this theory, *every body is continuously emitting radiant energy in all directions at a rate depending only on the nature of its surface and its temperature, and it is absorbing radiant energy from all surrounding bodies at a rate depending on its surface and the temperature of the surrounding bodies* .

If a hot body *A* is placed in an evacuated enclosure *B* , at a lower temperature than *A* , then *A* emits more energy per unit time than that it absorbs. Hence it cools until it reaches the temperature of *B*. Similarly, if a body *C*, cooler than *B*, is put in *B*, then *C* warms upto the temperature of *B*.

The exchange of energy continues even after *A* and C reach the temperature of *B*, but then each *A* and *C* emits as much energy per unit time as it absorbs and hence its temperature remains constant. Thus the temperature-equilibrium between a body and its surroundings is a dynamic equilibrium ; the body emits radiant energy to its surroundings at the same rate at which it absorbs energy from the surroundings.

A Body Placed in an Enclosure: Let us consider an enclosure made of non-conducting walls and maintained at a temperature T. The walls of the enclosure continually emit and absorb radiant energy so that the enclosure is filled with radiation. Let a body at a temperature T' be placed inside the enclosure. According to Prevost's theory, there will be an exchange of radiation between the walls of the enclosure and the body until an equilibrium is attained when each emits radiant energy at the same rate at which it absorbs.

Suppose, at equilibrium, $T' < T$. Then a heat engine can be used to absorb some heat from the walls of the enclosure, convert a part of it into work, and reject the rest to the body. This can be continued until $T' = T$. According to our assumption, the system is no more in equilibrium. If left, it would again be in equilibrium with $T' < T$. The difference in temperatures can again be utilised to obtain work. Thus, an unlimited amount of work can be obtained without expenditure of energy, which is a violation of the second law of thermodynamics. Hence our supposition that at equilibrium $T' < T$ is wrong. At equilibrium we must have $T' = T$. That is, *the body placed in the enclosure would acquire the temperature of the enclosure.*

Radiation Inside the Enclosure is Homogeneous and Isotropic : The rate of energy emission from a body depends only on the nature of the surface of the body and its temperature. Hence it would remain the same if the body changes either its position inside the enclosure or its orientation.

The body would also absorb radiant energy at the same rate *for all its positions and orientations* , because at equilibrium a body absorbs at the same rate at which it emits. This is possible only when the intensity of radiation inside the enclosure is same at all points and in all directions, because the amount of radiation absorbed by the body depends on the intensity of radiation. Hence, *the radiation inside the enclosure is homogeneous and isotropic* .

Radiation Dependent on Temperature Only : Let us consider two enclosures E_1 and E_2 of walls made of different materials but maintained at the same constant temperature. Each enclosure is filled with radiant energy. Suppose that the density of the radiant energy is greater in E_1 than in E_2 .

Let the two enclosures be joined to each other by a screen transparent to the radiation. Naturally, some radiation will be transmitted through the screen from E_1 to E_2. Now, let the enclosures be separated and allowed to regain equilibrium with their new radiation-densities. Since the energy-density in E_1 is now smaller than that corresponding to its original temperature, its walls will emit energy to the space inside it and so their temperature will fall. Similarly, the temperature of the walls of E_2 will rise. Thus a difference of temperature will be set up between E_2 and E_1 . Now E_2 can be used as the source and E_1 as the sink in a heat engine and work can be obtained *without expenditure of energy* , which is a violation of the second law of thermodynamics. Therefore, our supposition that the radiation-density in E_1 is greater than in E_2 is wrong. It must be same in E_1 and E_2 (which are at the same temperature). This is true for each small band of wave-lengths present in the radiation because we can give same argument by putting a screen transparent to each band of wave-lengths. Thus, *the quantity and quality of the radiation energy in a uniform temperature enclosure depend only on the temperature of the enclosure, and are independent of the nature of the walls* .

Let us now suppose that a number of bodies *A, B, C, D* of different natures are placed inside an enclosure. Each body will finally attain the temperature of the enclosure.

Suppose *A* is black, *B* is highly polished, *C* is transparent to the radiation, and *D* can absorb only radiation of one particular wave-length. *A* will absorb most of the radiation falling on it. As its temperature is constant, it follows from the theory of exchanges that it will emit radiation at the same rate as it absorbs. *B* absorbs very little radiation and so it will emit radiation at the same small rate as it absorbs. C transmits all the radiation falling on it and thus it will not affect the balance of radiation in the enclosure. *D* can absorb radiation only of a particular wave-length and so it will emit that particular radiation at an equal rate. Thus , each body emits radiant energy exactly at the same rate and exactly of the same kind as it absorbs. Thus, the bodies present in the enclosure do not affect the quantity or quality of the radiant energy in it.

Therefore, *the quantity and quality of the radiation inside a uniform temperature enclosure is independent of the nature of the wall of the enclosure and of the nature of the bodies present in it, and depend only on the temperature of the enclosure* . It is therefore known as "temperature radiation".

Radiation Inside Enclosure is Identical with Black-body Radiation : Since the radiation inside an enclosure is independent of the presence of any body inside the enclosure, a perfectly black-body placed in the enclosure would emit an energy of the same quality and quantity as it would absorb. This is possible only when the radiation inside the enclosure is identical in every respect with that emitted by a black-body. It is, therefore, also known as "black-body radiation" or "full radiation."

Bartoli's Proof of the Existence of Pressure of Radiation: Maxwell proved from the electromagnetic theory that the radiation should exert a pressure, and this is equal to the energy density of radiation for a parallel beam. A thermodynamic proof of the existence of radiation pressure is due to Bartoli. This is as below:

Let us 'imagine' a cylinder with perfectly reflecting walls, and its end-faces closed by perfectly conducting 'black' pieces *D* and C. Let D and C be maintained at constant temperatures

T_1 and T_2 (by keeping them in contact with heat sources of infinite capacity and temperatures T_1 and T_2); T_1 *being greater than* T_2 . Let A and B be two perfectly reflecting screens; and V be a valve in the screen B .

Let the initial positions of the screens A and B be as shown, the valve V being open. The space AC is filled with radiation which is in equilibrium with black body C at temperature T_2; whereas the space AD is filled with radiation which is in equilibrium with black body D at temperature T_1 .

Let us now imagine that the valve V is closed and the screen B is pushed up to the position B'. The radiation, initially in AB , is compressed within the space AB' and its density in AB' becomes greater than that in AD . If now the screen A is withdrawn, the space $B'D$ will have much more radiation than necessary for equilibrium with the black body D at temperature T_1 .

The excess radiation will therefore be absorbed by D . Thus we have transferred heat from a colder body C to a hotter body D . This, by the second law of thermodynamics, could be done by the expenditure of the work. This means that we have done some work in pushing the screen B upward, *i.e.*, B has been pushed upwards against a pressure. This pressure can be due only to radiation.

Pressure of Diffuse Radiation : Let us consider radiation of energy-density u (amount of radiation per unit volume) falling normally upon a surface. Radiation travels with the speed of light c . Therefore, the radiation reaching unit surface area per second

$$= cu,$$

which is the amount of radiation contained in a cylinder of length c and cross-sectional area unity.

We know that with energy u is associated a momentum

u/c. Therefore, the momentum reaching surface area per second is $cu/c = u$. Since the rate of change of momentum per unit surface area is the pressure, the surface absorbing the radiation falling upon it experiences a pressure given by

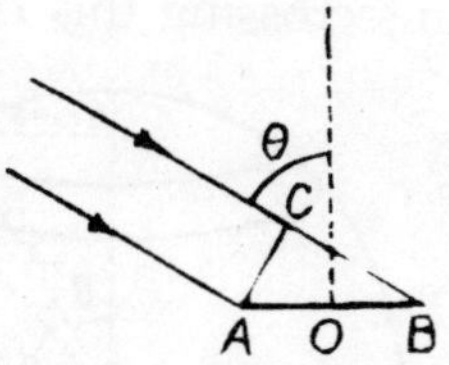

$$p = u$$

Now, suppose the radiation is falling on a surface AB at an angle θ with the normal. The energy contained in a cylinder of length c and cross-sectional area unity now falls on an area $\frac{1}{\cos\theta}\left[\because \frac{AB}{AC} = \frac{1}{\cos\theta}\right]$. The force per unit area on the surface in the direction of falling radiation is now

$$\frac{u}{1/\cos\theta} = u\cos\theta$$

Its normal component is

$$(u\cos\theta)\cos\theta = u\cos^2\theta$$

If this radiation is absorbed, the pressure experienced by the surface is

$$p = u\cos^2\theta$$

Let us now consider radiation in a uniformly-heated enclosure. This can be considered as equivalent to a large number of beams, say N, of equal intensity distributed uniformly in all directions. The pressure experienced by a surface AB in such an enclosure would be

$$p = \Sigma Nu\cos^2\theta = Nu\Sigma\cos^2\theta \quad \ldots\text{(i)}$$

where $\Sigma\cos^2\theta$ is the sum of the values of $\cos^2 0$ for all the beams.

To evaluate $\Sigma\cos^2\theta$, we draw a hemisphere of radius r around the centre O of the surface AB. On this hemisphere we take a ring of radius r sin θ and width $r\,d\theta$, as cut by two cones of semi-angles θ and $\theta + d\theta$ drawn from O as apex. If dN is the number of beams crossing this ring, then

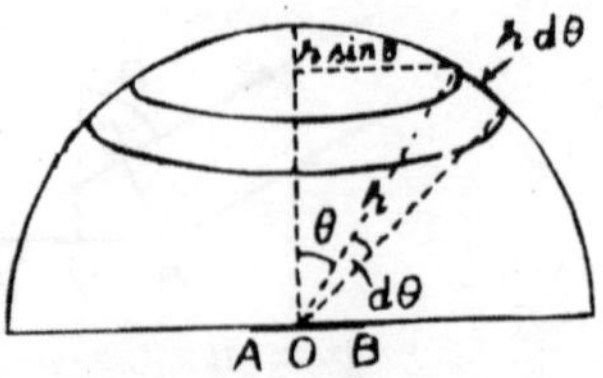

$$\frac{dN}{N} = \frac{\text{area of ring}}{\text{area of hemisphere}}$$

$$= \frac{2\pi(r\sin\theta)r\,d\theta}{2\pi r^2} = \sin\theta\,d\theta$$

or $$dn = n\ Sin\ \theta\ d\theta$$

The value of $\cos^2\theta$ is same for all the dN beams crossing the ring. Thus, for the ring, we have

$$\Sigma\cos^2\theta = \cos^2\theta\,dN.$$

and so for the entire hemisphere, we have

$$\Sigma\cos^2\theta = \int\cos^2\theta\,dN = N\int_0^{\pi/2}\cos^2\theta\sin\theta\,d\theta = \frac{1}{3}N$$

Substituting this value in eq. (i), we get

$$\boxed{p = \frac{u}{3}}$$

Thus, the pressure due to diffuse radiation is one-third the energy-density.

Analogy between Black-body Radiation and Perfect Gas: There is an analogy between black-body radiation *i.e.* radiation in an enclosure and perfect gas. A perfect gas is an assembly of molecules having all velocities from 0 to ∞ and moving in all directions. In the same way, radiation in an enclosure consists of all wave-lengths and is proceeding in all directions. In a gas, the molecules have Maxwellian velocity distribution which shows marked resemblance with the energy distribution among the wave-lengths of the black-body radiation. The gas molecules and the radiation both exchange momentum with the walls of the enclosure and exert pressure on them.

Black Body : A perfectly black body is one which absorbs completely all the radiation, of whatever wave-length, incident on it. Since it neither reflects nor transmits any radiation, it appears black whatever the colour of the incident radiation may be.

If a black body be placed inside a uniform temperature enclosure, it will absorb the full radiation of the enclosure. Since the quantity and quality of the radiation inside a uniform temperature enclosure is not affected by the presence of any body inside it, the black body will emit the full radiation of the enclosure on attaining temperature equilibrium with it. Therefore, the radiation emitted by a black body is also the 'full radiation' consisting of all possible wave-lengths.

Realization of Black Body in Practice : There is no surface available in practice which will absorb all the radiation falling on it. Even the lamp-blacked surface which absorbs practically all the visible and infra-red radiation, reflects the far infra-red radiation.

Fery has, however, designed a body which approximates very closely to the properties of a perfectly black body. This body consists of a hollow copper sphere blackened on the inside and having a small hole *O* in the surface. Any radiation falling on the hole will suffer complete absorption at the interior surface. The fraction of the radiation of any particular wave-length not absorbed at its first striking the blackened surface will ultimately be absorbed by successive reflections at this surface. Only a very small portion of the radiation, depending upon the size of the hole, will come back out of the hole. To eliminate direct reflection of radiation from the interior surface just opposite the hole *O*, a conical projection *P* is made. The small *hole O* thus behaves very closely like a black body. If the sphere be heated, the radiation coming out of the hole will be very nearly the black-body radiation.

Wien designed another type of black body. It is a long metallic tube *C* blackened inside and surrounded by concentric porcelain tubes *P*, *P'*. It is heated

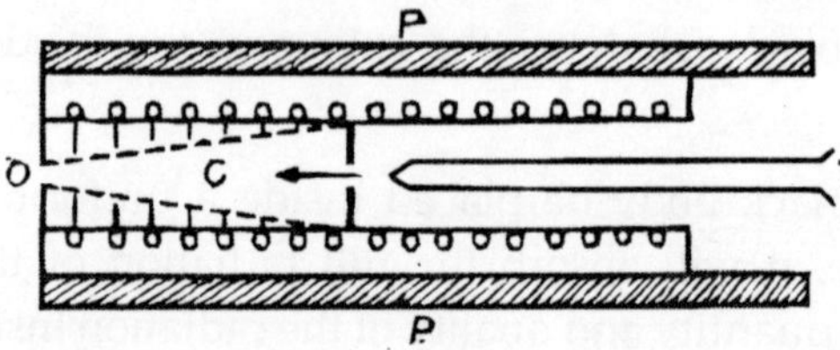

by an electrical current flowing in a coil wound around it. The radiation passes through a number of limiting diaphragms and comes out through a narrow hole *O* in the wall of the tube. The temperature of the central part of the tube can be measured by a thermocouple *T*.

Emissive Power

The emissive power of a body, at a given temperature and for a given wave-length, is defined as the radiant energy emitted per second per unit surface area of the body per unit wave-length range. Thus, if $e_\lambda \; d\lambda$ be the radiant energy between the wave-lengths λ and $\lambda + d\lambda$, emitted per second per unit surface area of a body of a temperature T, then in the limit $d\lambda \to 0, e_\lambda$ will be the "emissive power " of the body at temperature *T* for the wave-length λ.

In fact, the above is the definition of the "monochromatic" or "spectral" emissive power e_λ. The total emissive power *e* (say) is the radiant energy emitted per second per unit surface area at *all* wave-lengths.

Absorptive Power

The absorptive power of a body, at a given temperature and for a given wave-length, is defined as the ratio of the radiant energy absorbed per second by the surface of the body to the total energy falling per second on the same area. Thus, if an amount dQ of radiant energy between wave-lengths λ and $\lambda + d\lambda$ be falling on a body at a temperature T, of which a fraction $a_\lambda \, dQ$ be absorbed, then, in the limit $d\lambda \to 0, a_\lambda$ will be the "absorptive power" of the body at temperature T for the wave-length λ.

Again, the above is the definition of the "monochromatic" or "spectral" absorptive power a_λ. The total absorptive power a (say) is the ratio of the energy absorbed per second to the energy falling per second at *all* wave-lengths.

Kirchhoff's Law

It states that *the ratio of the emissive power to the absorptive power for radiation of a given wave-length is the same for all bodies at the same temperature, and is equal to the emissive power of a perfectly black body at that temperature.*

Proof: Let a body be placed in a uniformly-heated enclosure at a constant temperature. Let an amount of radiant energy dQ between the wave-lengths λ and $\lambda + d\lambda$ be incident per second on unit surface area of the body. Let a_λ be the absorptive power of the body for the wave-length λ. Then, an amount of energy $a_\lambda \, dQ$ will be absorbed per second by unit surface area of the body. The balance $(1 - a_\lambda) dQ$ of the incident energy will be reflected or transmitted.

Now, let e_λ be the emissive power of the body at wave-length λ. Then an amount of energy $e_\lambda \, d\lambda$ between the wave-lengths λ and $\lambda + d\lambda$ will be emitted per second by unit surface area of the body by virtue of its temperature.

Thus, the total energy sent out by unit area of the body per second is $(1-a_\lambda)dQ + e_\lambda \, d\lambda$.

Now, the presence of the body does not effect the quantity or quality of the radiation stream in the enclosure. Therefore, the energy sent out by unit area of the body per second should be equal to the energy received, and hence

$$(1-a_\lambda)dQ + e_\lambda \, d\lambda = dQ$$

or
$$e_\lambda \, d\lambda = a_\lambda dQ \qquad \ldots\text{(i)}$$

Now, for a perfectly black body the absorptive power $a_\lambda =$ 1 (for all wave-lengths). Therefore, if E_λ be the emissive power of a black body, we shall have

$$E_\lambda d\lambda = dQ$$

Substituting this value of dQ in eq. (i), we get

$$e_\lambda d\lambda = a_\lambda E_\lambda d\lambda$$

or
$$\frac{e_\lambda}{a_\lambda} = E_\lambda$$

Since E_λ is constant at a given temperature, it follows that

$$\frac{e_\lambda}{a_\lambda} = \text{constant},$$

for all substances at the same temperature. This is Kirchhoff's law.

Illustrations and Applications: Kirchhoff's law tells us *that good absorbers are good emitters. If a body absorbs radiation of a particular wave-length strongly, it also emits the same radiation strongly.*

When a polished metal ball having a black spot on its surface is heated to a high temperature, the black spot appears more brightly than the polished surface. The reason is that the black spot, which absorbs light more strongly than the polished surface at ordinary temperature, emits more strongly at high temperature. Similarly, a piece of red glass, which absorbs green light strongly, glows with a green light when heated.

The best example of Kirchhoff's law is provided by the emission and absorption spectra of sodium vapour. When the sodium vapour is heated to a temperature, high enough to emit visible radiation, its emission spectrum consists of two yellow lines. On the other hand, when white light is passed through a cooler sodium vapour and then seen through a spectroscope, the continuous spectrum of the white light is found to consist of two dark lines in exactly the same positions as the yellow lines in the emission spectrum. Thus, sodium vapour which emits two yellow lines strongly is also a good absorber of light of these two wave-lengths. The phenomenon enabled Kirchhoff to explain the *Fraunhofer dark lines* crossing the continuous spectrum of the sun. He explained that the sun consists of a central glowing mass which gives the continuous spectrum. It is surrounded by a cooler atmosphere which contains various elements like hydrogen, nitrogen, sodium, calcium, copper, etc. in the gaseous state. When radiation from the central mass passes through the surrounding atmosphere, the various elements absorb those wave-lengths which they can emit at a higher temperature. As a result, those wave-lengths are missing from the sun spectrum and we see dark lines in their places.

Kirchhoff's law is responsible for the development of the science of spectrum analysis. It established that the atoms of each element give a spectrum which is characteristic of that element alone. Hence , if the characteristic spectral lines of an element are seen in a spectrum, it is certain that the element is present in the substance of which the spectrum is being taken.

Stefan's Law : It states that ***the total radiant energy E emitted per second from unit surface area of a black body is proportional to the fourth power of its absolute temperature T.***

Thus

$$E \propto T^4$$

or

$$E = \sigma T^4$$

where σ is a constant, called "Stefan's constant". Boltzmann established this law theoretically from thermodynamic considerations. Hence it is usually called as 'Stefan-Boltzmann law.'

Now, a black body at absolute temperature T surrounded by another black body at absolute temperature T_0 will loose an amount of energy σT^4 per second per unit area, and gain from the surroundings an amount σT_0^4 per second per unit area. Hence the *net* loss of energy per second per unit area of the body will be given by

$$E = \sigma\left(T^4 - T_0^4\right)$$

Thermodynamic Derivation of Stefan's Law : Let us consider a cylinder of perfectly reflecting walls and fitted with a perfectly reflecting piston. Suppose it is filled with black-body radiation which is in equilibrium with the interior walls. The energy density of radiation u (ratio of energy to volume) is independent of the material of the walls and depends on the temperature T only. Now, according to Maxwell's electromagnetic theory, the black-body radiation in an enclosure exerts a pressure p on the walls which is equal to one-third of the energy-density. That is

$$p = \frac{1}{3}u$$

Thus, the black-body radiation is completely described by three coordinates p (pressure of radiation), V (volume of radiation) and T (temperature of walls with which the radiation is in equilibrium). Hence it may by treated as a 'thermodynamic system'.

Let us now imagine that an infinitesimal amount of heat dQ flows into the cylinder from outside, and at the same time the volume changes from V to $V + dV$. Let dU be the change in the internal energy of radiation, dS the change in its entropy, and dW the external work done by the radiation in expanding through dV. Then the I and II laws of thermodynamics give

$$dQ = dU + dW = dU + p\,dV$$

and $$dQ = T\,dS.$$

These give $$dU + p\,dV = T\,dS$$

or $$dS = \frac{dU + p dV}{T}$$

Making this substitution in Maxwell's second relation $\left(\frac{\partial S}{\partial V}\right)_T = \left(\frac{\partial p}{\partial T}\right)_V$, we get

$$\frac{1}{T}\left(\frac{\partial U + p\partial V}{\partial V}\right)_T = \left(\frac{\partial p}{\partial T}\right)_V$$

or $$\left(\frac{\partial U}{\partial V}\right)_T + p = T\left(\frac{\partial p}{\partial T}\right)_V$$

Now, $U = Vu$ and $p = u/3$, where u is a function of T only. Thus

$$\left(\frac{\partial (Vu)}{\partial V}\right)_T + \frac{u}{3} = \frac{T}{3}\left(\frac{\partial u}{\partial T}\right)_V$$

or $$u + \frac{u}{3} = \frac{T}{3}\left(\frac{\partial u}{\partial T}\right)$$

or $$\frac{4u}{3} = \frac{T}{3}\left(\frac{du}{dT}\right)$$

or $$\frac{du}{u} = 4\frac{dT}{T}$$

Integrating: $$\log u = 4 \log T + a$$

or $$u = aT^4$$

where a is the integration constant.

We can show that the total energy E radiated per second from unit area of a perfectly black body at absolute temperature T and the energy-density of radiation u inside an enclosure at the same temperature are related by the equation

$$E = \frac{1}{4}uc$$

where c is the velocity of light. Therefore, substituting the value of u from the last expression, we get

$$E = \frac{1}{4}acT^4$$

or $$E = \sigma T^4$$

where $\sigma = \frac{1}{4}ac$ and is called "Stefan's constant." This is Stefan's law.

Experimental Verification of Stefan's Law : Lummer and Pringsheim verified the Stefan's law between 100° and 1300 °C.

They took two sources of black-body radiation, A and B. The body A was a double-walled copper vessel blackened inside and having boiling water

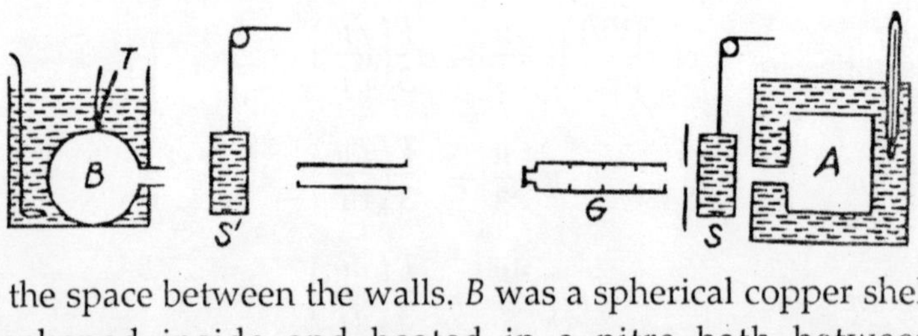

in the space between the walls. B was a spherical copper shell blackened inside and heated in a nitre bath between temperatures 100° and 600 °C. Its temperature was measured by a thermocouple T. (For higher temperatures, an iron cylinder blackened inside and heated in a double-walled gas furnace was used). Water-cooled shutters *S* and *S'* were placed infront *A* and *B*. A Lummer and Kurlbaum bolometer *G* was used to measure the radiant energy. It could be moved on a scale.

First of all, the shutter *S* was raised so that the bolometer *G* received the radiation from the black body *A* . The deflection in the galvanometer of the bolometer was noted. This was repeated by putting *G* at various distances from *A*. The deflection was found to be inversely proportional to the square of this distance. This indicated that the deflection was proportional to the energy of the radiation falling on the bolometer.

The nitre bath was then heated to a steady temperature *T*, and the shutter *S′* was raised. Thus the radiation from *B* fell on the bolometer. The deflection of the galvanometer, *d* , was read. Let T_0 be the temperature of the shutter *S′*. According to the inverse-square law, the radiant energy received by the bolometer from the body *B* is the same as would have been if the temperature of the shutter *S′* itself were raised from T_0 to *T*. Thus, according to Stefan's law, the additional radiant energy which fell on the bolometer when the shutter *S′* was raised would be proportional to $(T^4 - T_0^4)$. Then, we must have

$$d \propto (T^4 - T_0^4).$$

Lummer and Pringsheim kept the temperature of *S′* at 290 K *(i.e.* T_0 = 290 K). The body *B* was raised to different temperatures, *T*, and corresponding deflections, *d*, were read. The deflection *d* was always found to vary as

$$d \propto (T^4 - 290^4)$$

Thus the Stefan's law was verified.

Stefan-Boltzmann Law : According to Stefan-Boltzmann law, the radiant energy (heat) lost per second per unit area by a black body at absolute temperature T, surrounded by an enclosure at temperature T_0, is given by

$$E = \sigma\left(T^4 - T_0^4\right)$$

Let us suppose that T is greater than T_0 by only a small amount x. Then we can put $T = (T_0 + x)$. Then

$$E = \sigma\left[\left(T_0 + x^4\right) - T_0^4\right]$$

$$= \sigma\left[4T_0^3 x + 6T_0^2 x^2 + 4T_0 x^3 + x^4\right]$$

Now x is small say 30° whereas T_0 is 300 K for a room temperature of 27 °C (say). Therefore, the terms $6T_0^2x^2$, $4T_0x^3$, and x^4 can all be neglected compared with $4T_0^3x$. Thus , the last expression reduces to

$$E = 4\sigma T_0^3 x = 4\sigma T_0^3 (T - T_0)$$

or

$$E \propto (T - T_0)$$

Thus, when the temperature of a body is only *slightly higher* than the temperature of its surroundings, the heat lost per second by the body is proportional to the difference in temperature of the body and that of the surroundings. This is Newton's law of cooling.

Determination of Stefan's Constant in Laboratory : The apparatus consists of a hollow metallic hemisphere *H* blackened inside and placed in a wooden box *W*, which is lined with tin. It serves as a steam chamber. The whole arrangement is placed symmetrically on a platform *AB* which has a small hole at its centre. A small silver disc *D* blackened on its top surface can be fitted in or taken out from the hole. One junction of a silver-constant thermocouple is soldered to the lower surface of *D*, while the other is placed in a test-tube containing oil kept in a water-bath C.

Theory : When the inner surface of *H* is heated, it acts as a black-body radiator. The disc *D* absorbs the radiation emitted by *H* and its temperature rises continuously.

Let T_1 be the steady absolute temperature of *H*, and T_0 the temperature of the disc *D* when it is just exposed to radiation from *H*. If E_1 be the radiant energy absorbed by the disc per second per unit area and E_0 that emitted by it, then by Stefan's law

$$E_1 = \sigma T_1^4$$

and $$E_0 = \sigma T_0^4$$

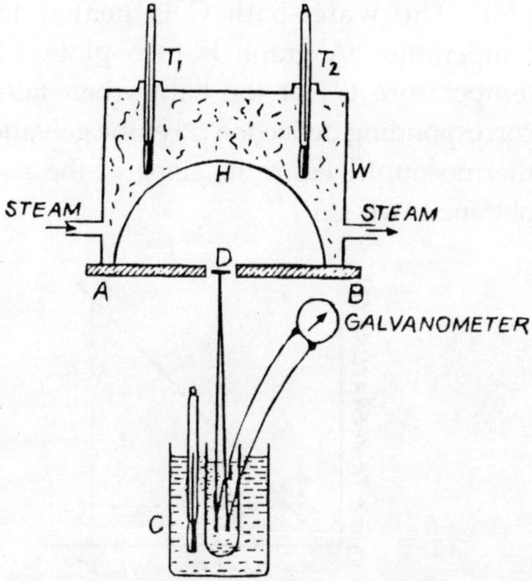

where σ is Stefan's constant. Thus, if *A* cm² be the area of the disc, the net energy gained per second by the disc

$$= (E_1 - E_0)\ A\ erg$$

$$= \sigma\left(T_1^4 - T_0^4\right) A \text{ erg}$$

$$= \frac{\sigma\left(T_1^4 - T_0^4\right) A}{J} \text{calorie}$$

where *J* is number of ergs per calorie. Bur if *m* gm be the mass of the disc, *s* its specific heat and *dT/dt* its rate of rise of temperature at T_0, then the energy gained per second by it will be *ms (dT/dt)*. Hence

$$ms\frac{dT}{dt} = \frac{\sigma\left(T_1^4 - T_0^4\right)A}{J}$$

or
$$\sigma = \frac{Jms}{A\left(T_1^4 - T_0^4\right)}\left(\frac{dT}{dt}\right). \qquad \text{...(i)}$$

Procedure : The experiment is performed in two steps :

(i) *Calibration of the Thermocouple* : Before passing steam in the chamber *W*, the silver disc is kept in position in the hole. The water-bath C is heated to *a* suitable temperature. A graph is then plotted between the temperature (*T*) of the bath when falling, and the corresponding deflection *(x)* of the galvanometer of the thermocouple. From the slope of the straight line so obtained, we get

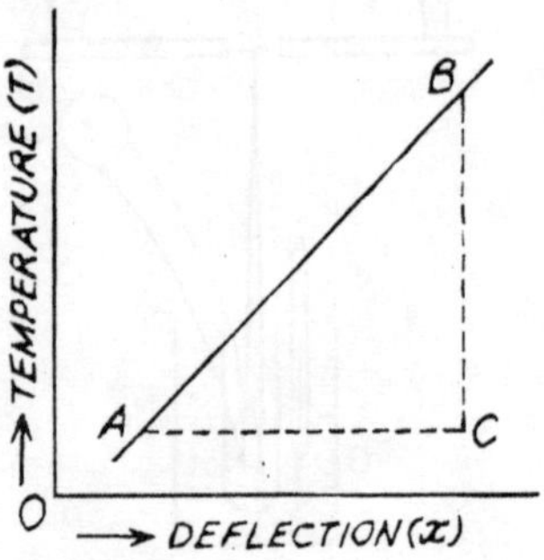

$$\frac{dT}{dx} = \frac{BC}{AC} \qquad \text{...(ii)}$$

The disc has acted as the cold junction of the thermocouple.

(ii) *Determination of (dT/dt)* : The disc is removed from the hole and the steam is passed through the box *W*. When the temperature becomes steady, the disc is again placed. It now absorbs radiation from *H* and its temperature rises. Therefore, the deflection in the galvanometer increases. A graph between the deflection *(x)* and time *(t)* is plotted. Now, a tangent is drawn at a point *P* taken as low as possible on the curve. From the slope of the tangent at this point, we have

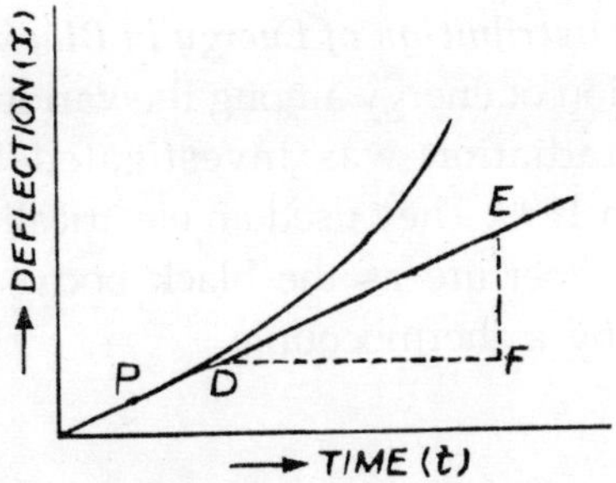

$$\frac{dx}{dt} = \frac{EF}{DF} \qquad \text{...(iii)}$$

Equations (ii) and (iii) give

$$\frac{dT}{dt} = \frac{dT}{dx} \times \frac{dx}{dt} = \frac{BC}{AC} \times \frac{EF}{DF}$$

Thus, the value of dT/dt at temperature corresponding to the point P is obtained. m, s and A are constants for the disc. T_1 is the steady temperature of the steam chamber. T_0, the temperature corresponding to the point P, is the temperature of the disc when it is just exposed to radiation. It may be, therefore, taken as the room temperature. Hence the Stefan's constant σ is readily obtained from eq. (i).

A body, irrespective of its colour or any other physical characteristic, behaves as a perfectly black body when placed in a uniformly-heated enclosure and is in thermal equilibrium with it. Therefore, both the glass pieces will appear black.

The piece inside the furnace will appear brighter because it behaves as a perfectly black body and hence is a better emitter. The piece taken outside is no more a black body.

The hole behaves like a black body. So the radiancy of the hole, given by σT^4, is larger than the radiancy of the outer surface of the jug given by $e\sigma T^4$, where e is the surface emissivity and is less than 1. Therefore, the radiation coming out from the hole is more intense than that coming from the surface of the jug, although both are at the same temperature. Hence the hole appears brighter than the surface.

Spectral Distribution of Energy in Black-body Radiation : The distribution of energy among the various wave-lengths in black-body radiation was investigated by Lummer and pringsheim in 1899. They used an electrically-heated chamber with a small aperture as the black body and measured its temperature by a thermocouple.

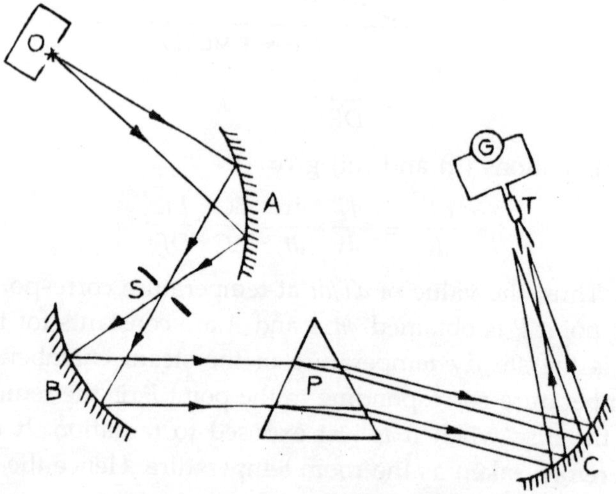

The radiation from the black body O fails on a slit 5 by means of a concave mirror A . The slit S is in the focal plane of a second concave mirror B which reflects the radiation as a parallel beam which is incident on a fluorspar prism P. The emerging beam falls on a concave mirror C which directs it on to a Lummer-Kurlbaum linear bolometer T. On rotating the mirror C about a vertical axis, the radiations of different wave-lengths fall on the bolometer one after the other. The deflection of the galvanometer gives the corresponding spectral radiancy E_λ, which is defined such that the quantity $E_\lambda d\lambda$ is the energy, for wave-lengths lying between λ and $\lambda + d\lambda$, emitted per second per unit surface area of the black body.

Fig. shows observed spectral radiancy E_λ plotted against the wave-length λ of radiation at three different black-body temperatures. These curves show three important features :

(i) The total energy emitted per second per unit area (*i.e.* radiancy E or area under the curve) increases rapidly with increasing temperature T. The increase is found in accordance with the Stefan's law :

$$E = \int_0^\infty E_\lambda = \sigma T^4$$

(ii) At a particular temperature, the spectral radiancy E_λ is a maximum at a particular wave-length λ_m (say). Most of the energy is emitted at wave-lengths not very different from λ_m.

(iii) The wave-length λ_m for maximum spectral radiancy decreases in direct proportion to the increase in temperature. This is called 'Wien's displacement law'. According to this law,

$$\lambda_m \times T = \text{constant.}$$

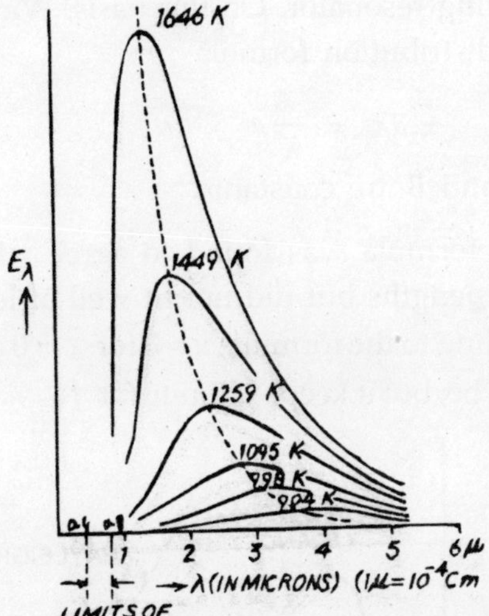

This means that as the temperature is raised, the cavity emits more and more radiation of the shorter wave-lengths.

Explanation by Classical Theory—Wien's and Rayleigh-Jeans Formulae: First of all, Wien in 1893, worked on the distribution of energy among the wave-lengths. He showed from pure thermodynamic reasoning that the spectral energy-density u_λ in the wave-length interval λ to $\lambda + d\lambda$ emitted by a black body at temperature T is of the form

$$u_\lambda d\lambda = \frac{A}{\lambda^5} f(\lambda T) d\lambda,$$

where A is a constant and $f(\lambda T)$, is an undetermined function. This is in agreement with the experimental result that the product λT at the peak of the $E_\lambda - \lambda$ curve is same at all temperatures.

In order to find the form $f(\lambda T)$, Wien assumed that the black-body radiation inside a cavity may be supposed to be emitted by resonators of molecular dimensions having Maxwellian velocity distribution, and the frequency of emitted radiation is proportional to the kinetic energy of the corresponding resonator. On this basis, Wien established the following distribution formula ;

$$u_\lambda d\lambda = \frac{A}{\lambda^5} e^{-B/\lambda T} d\lambda,$$

where A and B are constants.

Wien's formula was found to agree with experiment at short wave-lengths but did not fit well at long wave-lengths Fig. According to the formula, $u = 0$ for $\lambda = 0$ and also for $\lambda = \infty$ as it should be, but it keeps u_λ finite for $T = \infty$ which is unlikely.

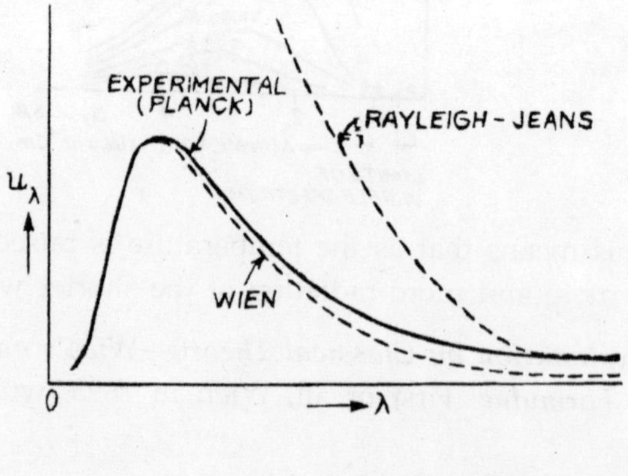

Rayleigh and Jeans considered the black-body radiator (cavity) full of electromagnetic waves of all wave-lengths between 0 and ∞ which, due to reflection at the walls, form standing waves. They calculated the number of possible waves having wave-lengths between λ and $\lambda + d\lambda$, and using law of equipartition of energy, established the following distribution formula :

$$u_\lambda d\lambda = \frac{8\pi kT}{\lambda^4} d\lambda,$$

where k is the Boltzmann's constant.

Rayleigh-Jeans formula was found to agree with experiment at long wave-lengths only. It is, however, open to a serious objection. As we move towards shorter wave-lengths (*i.e.* towards ultraviolet), the predicted energy would increase *without limit*, thus diverging enormously from experiment. This completely erroneous prediction is known as the 'ultraviolet catastrophe'.

Furthermore, at any temperature T the total energy of radiation $\int_0^\infty \frac{8\pi kT}{\lambda^4} d\lambda$, is predicted to be infinite which is against the Stefan's law.

Planck, in 1900, introduced entirely new ideas to explain the distribution of energy among the various wave-lengths of the cavity radiation and arrived at a formula which was in complete agreement with experiment at all wave-lengths.

Adiabatic Expansion of Black-body Radiation—Wien's Displacement Law : Let us consider a spherical enclosure with perfectly reflecting walls, and having a small black-body such as a grain of coal dust. In a short time the enclosure is filled with black-body radiation in equilibrium with the coal dust, at some temperature T (say). Let V be the volume of the enclosure and u the energy-density of radiation.

Let us imagine that the coal dust is taken out and the radiation is allowed to expand *adiabatically* by an outward motion of the walls of the enclosure. This would cause a decrease in the energy-density and fall in temperature. We can, however, argue that the black radiation, after adiabatic expansion, remains black radiation characteristic of the lower temperature.

For an adiabatic expansion ($dQ = 0$), the first law of thermodynamics gives

$$dU + p\,dV = 0$$

But $U = uV$ *and* $p = u/3$ *(by electromagnetic theory).*

$$\therefore \quad d(uV) + \frac{1}{3}u\,dV = 0$$

$$\text{or} \quad u\,dV + V\,du + \frac{1}{3}u\,dV = 0$$

$$\text{or} \quad \frac{4}{3}u\,dV + V\,du = 0$$

$$\text{or} \quad \frac{4}{3}\frac{dV}{V} + \frac{du}{u} = 0$$

On integration, we have

$$\frac{4}{3}\log_e V + \log_e u = \log_e C(\text{a constant})$$

$$\therefore \quad uV^{4/3} = C$$

But $u = aT^4$ *from Stefan's law, where a is a constant.*

$$\therefore \quad T^4V^{4/3} = \text{Constant}$$

$$\text{or} \quad \boxed{TV^{1/3} = \text{constant}}$$

Wien's Displacement Law : The radiation filled in the enclosure is undergoing continuous reflections from the moving walls of the enclosure. Due to Doppler's effect, a wave of wave-length λ which is incident on the wall at an angle θ suffers a change in wave-length given by

$$d\lambda = \frac{2v\cos\theta}{c}\lambda$$

where v is the velocity of the wall and c the velocity of radiation. Now, between two successive reflections say, at A and B, the wave travels a distance $2r\cos\theta$. Therefore, the number of reflections per second is $c/2r\cos\theta$, and that in a time dt is $\frac{c}{2r\cos\theta}dt$. Hence the change in wave-length in time dt is

$$d\lambda = \frac{2v\cos\theta}{c}\lambda \times \frac{c}{2r\cos\theta}dt = \frac{v\lambda}{r}dt$$

But the velocity of walls of the spherical enclosure of radius r is $v = \frac{dr}{dt}$, so that $vdt = dr$.

$$\therefore \quad \frac{d\lambda}{\lambda} = \frac{dr}{r}$$

Integration gives

$$\log_e \lambda = \log_e r + \log_e C \text{ (a constant)}$$

or $$\lambda = rC$$

or $$\frac{\lambda}{r} = \text{constant.} \qquad \text{...(ii)}$$

For adiabatic expansion, we have

$$TV^{1/3} = \text{constant.}$$

But $$V = \frac{4}{3}\pi r^3$$

$$\therefore \quad Tr = \text{constant.} \qquad \text{...(iii)}$$

Eq. (ii) and (iii) give

$$\boxed{\lambda T = \text{Constant} = \lambda' T'} \qquad \text{...(iv)}$$

This equation expresses the fact that if radiation of a particular wave-length at a certain temperature is adiabatically altered to another wave-length, then the temperature changes in the inverse ratio.

Let us now isolate waves of wave-lengths lying between λ and $\lambda + d\lambda$ in the spherical enclosure and allow these alone to undergo an adiabatic expansion. The spectral energy-density, the wave-length and the width of the wave-length band will be changed. Again, from the first law applied to the spectral components between λ and $\lambda + d\lambda$, we have

$$d(U_\lambda d\lambda) + pdV = 0.$$

For these components, we have

$$U_\lambda = u_\lambda V \text{ and } p = \frac{1}{3} u_\lambda d\lambda$$

$$\therefore \quad d(u_\lambda V d\lambda) + \frac{1}{3} u_\lambda d\lambda dV = 0$$

or $$d(u_\lambda d\lambda)V + u_\lambda d\lambda dV + \frac{1}{3} u_\lambda d\lambda dV = 0$$

or $$d(u_\lambda d\lambda)V + \frac{4}{3} u_\lambda d\lambda dV = 0$$

or $$\frac{d(u_\lambda d\lambda)}{u_\lambda d\lambda} = -\frac{4}{3}\frac{dV}{V} = -4\frac{dr}{r} \qquad \left[\begin{aligned} &\therefore V = \frac{4}{3}\pi r^3 \\ &\therefore \frac{dV}{V} = 3\frac{dr}{r} \end{aligned}\right]$$

Integration gives

$$\log_e (u_\lambda d\lambda) + 4 \log_e r = \log_e C(\text{constant})$$

or $$u_\lambda d\lambda r^4 = \text{constant}.$$

But from eq. (ii), $\lambda \propto r$ so that $d\lambda \propto dr$. Thus

$$u_\lambda r^4 dr = \text{constant}.$$

Integrating over *r*, we get $u_\lambda r^5$ = constant

But from eq. (iii), Tr = constant. Therefore, we get

$$u_\lambda T^{-5} = \text{constant.}$$

This, in view of eq. (iv), may also be written as

$$\boxed{u_\lambda \lambda^5 = \text{constant}}$$

This means that the energy-density changes inversely as λ^5 in adiabatic expansion.

Now, we know that spectral radiancy $E\lambda$ is proportional to energy-density u_λ. Therefore

$$\boxed{E_\lambda T^{-5} = \text{constant} = E_\lambda' T'^{-5}}$$

We can experimentally determine the distribution of energy (E_λ) among different wave-lengths (λ) at any temperature *T*, and draw the $(E_\lambda - \lambda)$ curve. From this curve, the curve for any higher temperature T' can be drawn by shortening the abscissa (λ) in the ratio

$$\frac{T}{T'}\left[\text{because} \frac{\lambda'}{\lambda} = \frac{T}{T'}\right]$$

and increasing the corresponding ordinate (E_λ) in the ratio

$$\left[\frac{T'}{T}\right]^5 \left[\text{because} \frac{E_\lambda'}{E_\lambda} = \left(\frac{T'}{T}\right)^5\right]$$

The wave-length-interval $d\lambda$ is also shortened in the ratio $\frac{T}{T'}$. The new curve thus obtained is higher but narrower, and the total area which represents the intensity is changed in the ratio.

$$\frac{T}{T'} \times \left(\frac{T'}{T}\right)^5 = \left(\frac{T'}{T}\right)^4$$

It is found that corresponding to peak point P in the T-curve there is a peak point P' in the T'-curve, such that

$$\boxed{\lambda_m T = \lambda_m' T' = \text{constant}}$$

where λ_m and λ_m' are the wave-lengths corresponding to maximum emission at T and T'' respectively. This is the usual form of Wien's displacement law.

Wien 's Distribution Law : We have shown that for the group of spectral components between λ and $\lambda + d\lambda$, we have

$$\lambda T = \text{constant}$$

and

$$u_\lambda \lambda^5 = \text{constant.}$$

Therefore, the spectral energy-density in the wave-length interval λ to $\lambda + d\lambda$ emitted by a black body at temperature T must be of the form

$$u_\lambda d\lambda = \frac{A}{\lambda^5} f(\lambda T) d\lambda$$

where A is a constant and $f(\lambda T)$ is an undetermined function of the product λT. This is the most common way of expressing Wien's distribution law.

It is not possible to determine the form of the function $f(\lambda T)$ by purely thermodynamic reasoning, but one has to make new assumptions regarding the nature of radiation. Wien assumed that the radiation inside the enclosure is produced by some kind of resonators of molecular dimensions and that the frequency of the waves emitted is proportional to the kinetic energy of the resonators. That is,

$$\frac{1}{2} mv^2 = av = ac/\lambda$$

where a is a constant and v is the frequency of the radiation emitted.

Now, according to Maxwell-Boltzmann distribution law,

the number of resonators having velocities between v and $v + dv$, or energies between av and $a(v + dv)$ is given by

$$N(v) = f(v)\ e^{-mv^2/2kT} = \phi(v)e^{-av/kT}$$

where $\phi(v)$ is a function of the energy of the molecular resonator. This, in terms of wave-length may be written as

$$N(\lambda) = \phi'(\lambda)e^{-ac/\lambda kT} = \phi'(\lambda)e^{-B/\lambda T}$$

where $B = ac/k$

The spectral energy-density in the wave-length interval λ to $\lambda + d\lambda$ may be taken to be proportional to $N(\lambda)$, so that

$$u_\lambda\, d\lambda = \phi'(\lambda)e^{-B/\lambda T}\, d\lambda.$$

Comparison with common form of Wien's law gives :

$$\phi'(\lambda) = \frac{A}{\lambda^5}$$

and $$f(\lambda T) = e^{-B/\lambda T}$$

Thus, we may write

$$u_\lambda d\lambda = \frac{A}{\lambda^5}e^{-B/\lambda T}d\lambda$$

where A and B are constants. This is Wien's distribution formula with two undetermined constants.

Modes of Vibration of Standing Waves in Black-body Radiation—Rayleigh-Jeans Law : Early in 1900, Rayleigh and Jeans presented a classical calculation for the energy-density of black-body radiation. They considered the radiation filled in an enclosure as consisting of standing electromagnetic waves of all possible frequencies. They computed the number of such (standing) waves in the frequency interval v to $v + dv$, and then applied the classical law of equipartition of energy to arrive at their energy distribution law.

To do this, let us consider a cubical enclosure of side l with perfectly reflecting walls, and let us place inside it a black particle. The radiation emitted by the particle is reflected by the walls and, as a result of interference between incident and reflected waves, the enclosure is filled with standing waves of various lengths, having nodal planes at the walls and at planes parallel to them and separated by $\lambda/2$. The 'allowed' wave-lengths parallel to the three edges of the cube, which form the three axes in space, are given by

$$l = n_x \frac{\lambda}{2}$$

$$l = n_y \frac{\lambda}{2}$$

and $$l = n_z \frac{\lambda}{2}$$

where n_x, n_y, n_z are positive integers (1, 2, 3......). But as the radiation is diffused, the waves can be inclined at any angle to the rectangular axes. For a wave making angles α, β, γ with the three axes, we would write

$$l\cos\alpha = n_x \frac{\lambda}{2},$$

$$l\cos\beta = n_y \frac{\lambda}{2},$$

and $$l\cos\gamma = n_z \frac{\lambda}{2},$$

where the direction cosines obey the relation

$$\cos^2\alpha + \cos^2\beta + \cos^2\gamma = 1$$

From the last four relations, we can write

$$n_x^2 + n_y^2 + n_z^2 = \left(\frac{2l}{\lambda}\right)^2$$

or $$n_x^2 + n_y^2 + n_z^2 = r^2 \qquad \text{...(i)}$$

where $$r = \frac{2l}{\lambda}.$$

Eq. (i) represents an ellipsoid with n_x, n_y, n_z as the coordinate axes. Any combination of positive integers *nx, ny, nz* satisfying eq. (i) will represent an allowed mode of vibration. Since n_x, n_y, n_z are 'integers', the possible combinations of them will be simply given by the number of points of intersection of the parallel lines drawn in space through successive values of n_x, n_y, n_z. These points clearly occur at the rate of one per unit volume, hence their number will be simply given by the volume of the ellipsoid. Further, since only "positive" values of n_x, n_y, n_z are permissible and these occur only in one out of the eight octants of the ellipsoid, the possible modes of vibration, f', are given by $\frac{1}{8}$th of the volume of the ellipsoid, that is,

$$f' = \frac{1}{8}\left(\frac{4}{3}\pi r^3\right)$$

$$= \frac{1}{8}\left(\frac{4}{3}\pi\right)\left(\frac{2l}{\lambda}\right)^3$$

$$= \frac{4\pi l^3}{3\lambda^3} = \frac{4\pi V}{3\lambda^3}$$

where V(= l^3) is the volume of the enclosure.

The number of modes of vibration (or of degrees of freedom) between λ and $\lambda + d\lambda$ is obtained by differentiating the above expression. That is,

$$df' = \frac{4\pi V}{\lambda^4} d\lambda$$

This number should be multiplied by 2 because transverse electromagnetic waves have two possible polarisations for each mode. Thus , the number of degrees of freedom of the waves is

$$df = \frac{8\pi V}{\lambda^4} d\lambda$$

The number *per unit volume* is clearly $(8\pi/\lambda)d\lambda$. Converting wave-length into frequency $v(=c/\lambda)$, the number becomes

$$df' = \frac{8\pi V}{c^3} \nu^2 d\nu$$

Hence the number of modes of vibration *per unit volume* in the frequency range ν to $\nu + d\nu$ is

$$\boxed{\frac{8\pi\nu^2 d\nu}{c^3}}$$

Now, each standing wave contains energy. According to the classical law of equipartition of energy, the average energy is the same for each wave in the enclosure, when the system is in thermal equilibrium, and is given by

$$\bar{\varepsilon} = kT$$

where k is the Boltzmann's constant and T is the temperature of the enclosure. Therefore, the energy-density inside the enclosure in the wave-length-range λ to $\lambda + d\lambda$ is

$$\begin{aligned} u_\lambda d\lambda &= \text{no. of waves } \textit{per unit volume} \times \text{average energy} \\ &= \frac{8\pi}{\lambda^4} d\lambda \times kT \\ &= \frac{8\pi kT}{\lambda^4} d\lambda \end{aligned}$$

This is Rayleigh-Jeans law.

In terms of frequency $\nu (= c/\lambda)$, we can write

$$u_\nu d\nu = \frac{8\pi kT\nu^2}{c^3} d\nu$$

The law agrees with experiment at long wave-lengths, but it fails miserably at short wave-lengths. For example, as $\lambda \to 0$, $u_\lambda \to \infty$, (ultraviolet catastrophy) whereas experimentally as $\lambda \to 0$, $u_\lambda \to 0$.

Planck's Mechanism of Emission and Absorption of Radiation: Planck in 1900, introduced entirely new ideas to explain the distribution of energy among the various wave-lengths of the cavity radiation. He assumed that *the atoms of* *the walls of the cavity radiator behave as oscillators, each with a characteristic frequency of oscillation.* These oscillators emit electromagnetic radiant energy into the cavity and also absorb the same from it, and maintain an equilibrium state. Planck made two rather revolutionary assumptions regarding these atomic oscillators :

(i) *An oscillator can have only discrete energies given by*

$$\in = nh\nu$$

where ν is the frequency of the oscillator, h is a constant known as 'Planck's constant', n is an integer known as 'quantum number'. This means that the oscillator can have only the energies $h\nu$, $2h\nu$, $3h\nu$, ... and not any energy in between. In other words, the energy of the oscillator is quantised.

(ii) *The oscillators do not emit or absorb energy continuously but only in 'jumps'.* That is, an oscillator emits or absorbs packets of energy, each packet carrying an amount of energy $h\nu$.

$$\Delta \in = (\Delta n)h\nu, \ \Delta n = 1, 2....$$

Average Energy of Planck's Oscillator : Let us now calculate the average energy of a Planck's oscillator of frequency ν. The relative probability that an oscillator has the energy $h\nu$ at temperature T is given by the Boltzmann factor $e^{-h\nu/kT}$. Now, let $N_0, N_1, N_2 ......N_r.....$ be the number of oscillators having the energies 0, $h\nu$, $2h\nu$,...$rh\nu$ ... respectively. Then, we have $N_r = N_0 e^{-rh\nu/kT}$.

The total number of oscillators is

$$N = N_0 + N_1 + N_2 + ...$$

$$= N_0\left(1 + e^{-h\nu/kT} + e^{-2h\nu/kT} + ...\right)$$

$$= \frac{N_0}{1 - e^{-h\nu/kT}} \qquad ...(i)$$

The total energy of the oscillators is given by

$$\begin{aligned}\in &= (N_0\times 0)+(N_1\times h\nu)+(N_2\times 2h\nu)+.....\\ &= (N_0\times 0)+\left(N_0e^{-h\nu/kT}\times h\nu\right)+\left(N_0e^{-2h\nu/kT}\times 2h\nu\right)+.....\\ &= N_0e^{-h\nu/kT}h\nu\left(1+2e^{-h\nu/kT}+3e^{-2h\nu/kT}+.....\right)\\ &= N_0e^{-h\nu/kT}\frac{h\nu}{\left(1-e^{-h\nu/kT}\right)^2} \qquad ...(ii)\end{aligned}$$

Dividing eq. (ii) by eq. (i), we obtain the *average* energy of an oscillator as given by

$$\bar{\in} = \frac{\in}{N}=\frac{e^{-h\nu/kT}(h\nu)}{1-e^{-h\nu/kT}}=\frac{h\nu}{e^{h\nu/kT}-1} \qquad ...(iii)$$

This is the expression for the average energy of a Planck's oscillator.

Planck's Radiation Formula

We can now combine Planck's expression for the average energy of an oscillator with the number of oscillators (or standing waves) per unit volume of the radiation as computed by Rayleigh and Jeans. Thus, the energy-density of radiation, u_ν, in the frequency range ν to $\nu+d\nu$ is given by

$$u_\nu d\nu = \frac{8\pi\nu^2 d\nu}{c^3}\times\bar{\in}$$

The term $\frac{8\pi\nu^2 d\nu}{c^3}$ represents the number of electromagnetic standing waves per unit volume in the frequency range ν to $\nu+d\nu$ in the cavity radiation.

Substituting the value of $\bar{\in}$ from eq. (iii), we have .

$$u_\nu d\nu = \frac{8\pi\nu^2 d\nu}{c^3}\left(\frac{h\nu}{e^{h\nu/kT}-1}\right)$$

or

$$\boxed{u_\nu d\nu = \frac{8\pi h\nu^3}{c^3}\frac{d\nu}{e^{h\nu/kT}-1}}$$

This is Planck's radiation formula in terms of frequency ν. To express it in terms of wave-length, we observe that, since $\nu = \frac{c}{\lambda}$,

$$d\nu = -\frac{c}{\lambda^2} d\lambda$$

and since an increase in frequency corresponds to a decrease in wave-length,

$$u_\lambda d\lambda = u_\nu d\nu$$

$$\text{Therefore } u_\lambda d\lambda = \frac{8\pi h}{c^3} \frac{\left(\frac{c}{\lambda}\right)^3 \left(\frac{c}{\lambda^2} d\lambda\right)}{e^{hc/\lambda kT} - 1}$$

or

$$\boxed{u_\lambda d\lambda = \frac{8\pi hc}{\lambda^5} \frac{d\lambda}{e^{hc/\lambda kT} - 1}} \qquad \text{...(iv)}$$

This is Planck's formula in terms of wave-length.

Explanation of Energy Distribution by Planck's Formula—Wein's Law and Rayleigh-Jeans Law are Special Cases : The Planck's formula is found to be in complete agreement with experiment for the entire wave-length range at all temperatures. This can be seen in the following way :

(1) When λ is very small, then $e^{hc/\lambda kT} >> 1$, so that Planck's formula given by eq. (iv) can be written as

$$u_\lambda d\lambda = \frac{8\pi hc}{\lambda^5} e^{-hc/\lambda kT} d\lambda$$

Putting $8\pi hc = A$

and $\frac{hc}{k} = B$, we have

$$u_\lambda d\lambda = \frac{A}{\lambda^5} e^{-B/\lambda T} d\lambda$$

This is Wein's law which agrees with experiment at short wave-lengths.

(2) When λ is very large, then $e^{hc/\lambda kT} \approx 1+\frac{hc}{\lambda kT}$, so that eq. (iv) can be written as

$$u_\lambda d\lambda = \frac{8\pi hc}{\lambda^5\left(1+\frac{hc}{\lambda kT}-1\right)}d\lambda = \frac{8\pi kT}{\lambda^4}d\lambda$$

This is Rayleigh-Jeans law which agrees with experiment at long wave-lengths.

Further, both Wien's displacement law and Stefan's law can also be derived from the Planck's formula.

The Planck's radiation formula

$$u_\nu d\nu = \frac{8\pi\nu^2 d\nu}{c^3}\left(\frac{h\nu}{e^{h\nu/kT}-1}\right)$$

has been deduced in the last question. The first term on the R.H.S. represents the number of resonators per unit volume lying in the frequency-range ν to $\nu+d\nu$, while the second term is the average energy of a Planck's resonator.

The plot of u_ν versus ν at temperatures T and $2T$ is shown in Fig.

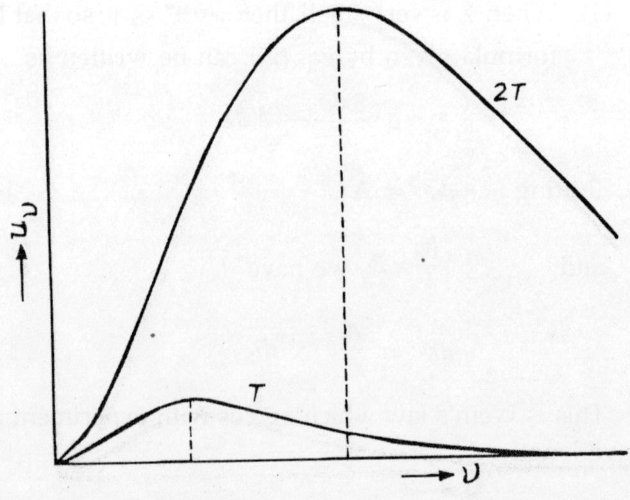

The area of $(u_\nu - \nu)$ curve on the ν-axis represents the *total* energy density which increases rapidly with temperature. In fact, the total energy density (area) at $2T$ is 16 times that at T (Stefan's law).

The frequency of maximum spectral energy density increases linearly with temperature (Wien's displacement law).

Wien's Displacement Law from Planck's Formula : The Planck's radiation formula is

$$u_\lambda d\lambda = \frac{8\pi hc}{\lambda^5}\frac{d\lambda}{e^{hc/\lambda kT} - 1}$$

or
$$u_\lambda = 8\pi hc\left(\lambda^{-5}\right)\left(e^{hc/\lambda kT} - 1\right)^{-1}$$

To find the wave-length at which the spectral radiancy is maximum, we put

$$\frac{du_\lambda}{d\lambda} = 0$$

that is

$$8\pi hc\left[-5\left(\lambda^{-6}\right)\left(e^{hc/\lambda kT} - 1\right)^{-1} + \lambda^{-5}(-1)\left(e^{hc/\lambda kT} - 1\right)^{-2} e^{hc/\lambda kT}\left(\frac{-hc}{\lambda^2 kt}\right)\right] = 0$$

or
$$\frac{5}{\lambda} = \left(e^{hc/\lambda kT} - 1\right)^{-1} e^{hc/\lambda kT}\frac{hc}{\lambda^2 kT}$$

or
$$5 = \frac{hc}{\lambda kT}\frac{e^{hc/\lambda kT}}{e^{hc/\lambda kT} - 1}$$

Putting $\frac{hc}{\lambda kT} = x$, we get

$$5 = \frac{xe^x}{e^x - 1} = \frac{x}{1 - e^{-x}}$$

or
$$\frac{x}{5} + e^{-x} = 1$$

This equation has a single root given by $x = 4.965$, and therefore x must be a constant. That is

$$\frac{hc}{\lambda kT} = 4.965.$$

Therefore, the wave-length λ_m at which the spectral radiancy per unit range of wave-length has its maximum value is given by

$$\lambda_m T = \frac{hc}{4.965k} = b(\text{say})$$

This is Wien's displacement law.

Stefan's Law from Planck's Formula : Let us write the Planck's radiation formula in terms of frequency :

$$u_\nu d_\nu = \frac{8\pi h}{c^3} \frac{\nu^3 d\nu}{e^{h\nu/kT} - 1}$$

The spectral radiancy E_ν, is related to the energy-density u_ν by

$$E_\nu = \frac{c}{4} u_\nu$$

Thus $$E_\nu d_\nu = \frac{2\pi h}{c^2} \frac{\nu^3 d\nu}{e^{h\nu/kT} - 1}$$

The total radiant energy over all frequencies is

$$E = \int_0^\infty E_\nu d_\nu = \frac{2\pi h}{c^2} \int_0^\infty \frac{\nu^3 d\nu}{e^{h\nu/kT} - 1}$$

Let us make the substitution $\frac{h\nu}{kT} = x$, so that $\nu = \frac{kT}{h} x$ and $d\nu = \frac{kT}{h} dx$. Then

$$E = \frac{2\pi k^4 T^4}{h^3 c^2} \int_0^\infty \frac{x^3 dx}{e^x - 1}$$

The value of the integral $\int_0^\infty \frac{x^3 dx}{e^x - 1}$ is $\frac{\pi^4}{15}$.

Thus $$E = \frac{2\pi^5 k^4}{15h^3c^2} T^4$$

Let us put $\frac{2\pi^5 k^4}{15h^3c^2} = \sigma$ (a universal constant). Then

$$E = \sigma T^4.$$

This is Stefan's law

Let us calculate the value of Stefan's constant from above:

$$\sigma = \frac{2\pi^5 k^4}{15h^3c^2}$$

$$= \frac{2\times(3.14)^5\times\left(1.38\times10^{-16}\,erg/K\right)^4}{15\times\left(6.63\times10^{-27}\,erg-\sec\right)^3\times\left(3\times10^{10}\,cm/\sec\right)^2}$$

$$= 5.6\times10^{-5}\,\mathrm{erg}/\left(\mathrm{cm}^2-\mathrm{sec}-\mathrm{K}^4\right)$$

Radiation Method for Temperature Measurement : To measure high temperatures by the method of radiation, two types of instruments are used :

(i) *Total Radiation Pyrometers* : In these instruments the total energy of the radiation emitted by the body under test is measured and the temperature is obtained by applying Stefan's law.

(ii) *Optical Pyrometers:* In these instruments the intensity of radiation of a certain wave-length emitted by the body is compared with that of the radiation of same wave-length emitted by a standard body at a known temperature, and its temperature is obtained by applying Wien's distribution law.

Fery's Total Radiation Pyrometer : Fery designed a total radiation pyrometer which is shown in Fig. It consists of a concave mirror C made of copper and nickeled on the front surface, having a small hole in the centre. It can be moved by a rack and pinion arrangement. Two plane semi-circular mirrors M, M inclined to each other at about 5° with an opening of about 1.5 mm at the centre are placed in front of the hole in the mirror C. The opening is backed by a blackened strip S. One junction of a thermocouple is soldered to the back of this strip and the other is protected from the direct radiation by a suitable screen (not shown). The thermo-couple has a millivoltmeter in its circuit.

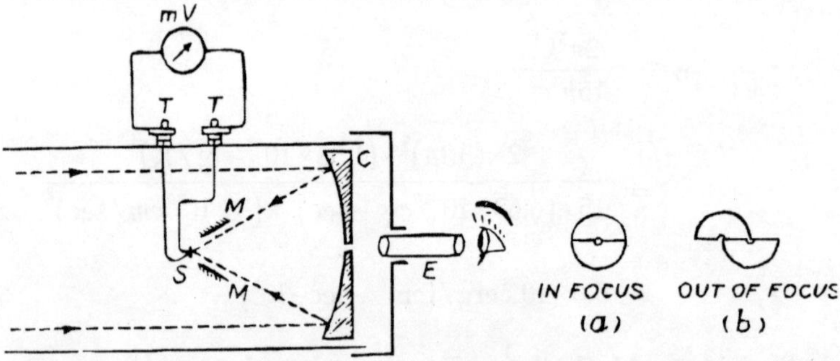

Working **:** The radiation beam from the hot body under test falls on the concave mirror C which focuses it on the opening between M, M. The opening is seen by an eye-piece E through the hole in the concave mirror C. The position of C is adjusted until the appearance through the eyepiece is as indicated in Fig. Thus the radiation, which is accurately focused on the opening between M, M falls on S, which is therefore heated and produces a thermo-emf, which is read by the millivoltmeter.

Calculation of Temperature **:** If T be the temperature of the hot body and T_0 that of the pyrometer case, then the reading of millivoltmeter will be given by

$$\theta = a(T^4 - T_0^4),$$

where a is a constant. If T is very much higher than T_0, then

$$\theta = aT^4$$

The instrument is usually calibrated by sighting it on a standard body which is raised to various temperatures. The temperatures are measured by a standard thermocouple and the corresponding readings of the millivoltmeter are noted. Then a calibration curve is drawn. The millivoltmeter may also directly be calibrated in degrees.

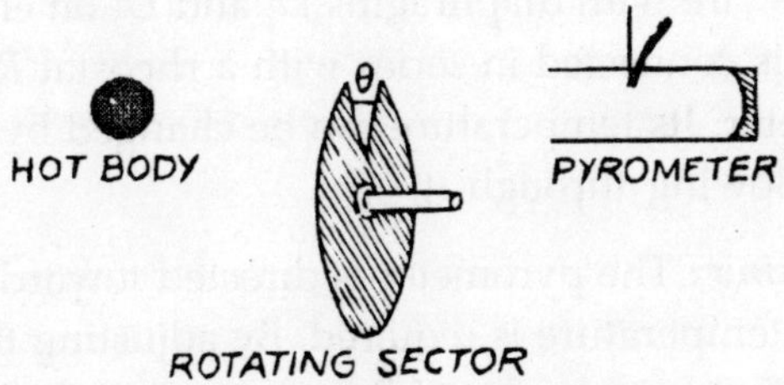

To measure temperatures above 1400 °C, a rapidly rotating sector is interposed between the hot body and the pyrometer. Thus if θ be the angle of sector, it will allow only a fraction θ /360 of the incident radiation to enter the pyrometer. The pyrometer then indicates a temperature which is less than the true temperature of the body. But the radiation is proportional to the fourth power of the absolute temperature. Therefore , if T be the true temperature of the body and T_1 that indicated by the pyrometer, then

$$T_1^4 = \frac{\theta}{360} T^4$$

$$T = Ti\left(\frac{360}{\theta}\right)^{1/4}$$

***Principle of an Optical Pyrometer*:** In an optical pyrometer, the intensity of radiation of a certain wave-length emitted by a hot body is compared with that of the radiation of the same wave-length emitted by a standard body. The instrument is adjusted until the two intensities are equal. The temperature of the hot body must then be equal to that of the standard body.

***Two Types of Optical Pyrometers are Popular* :** the disappearing filament pyrometer and the polarising pyrometer.

Disappearing Filament Pyrometer : A modified form of this pyrometer is shown in Fig. It consists of a telescopic objective *L* and an eyepiece *E* . An electric lamp *S* (standard body) having a straight filament is placed in the position of the cross-wire with diaphragms D_1 and D_2 on either side. The filament is connected in series with a rheostat *R*, battery, and an ammeter. Its temperature can be changed by changing the current flowing through it.

Working : The pyrometer is directed toward the hot body *B* whose temperature is required. By adjusting the position of the objective *L*, an image of *B* is focused on to the filament of *S*. The filament is seen through a red glass filter *F* by the eyepiece *E*. The temperature of *S* is adjusted by means of the rheostat *R*, until the filament disappears against the image of *B*. When the brightness of *S* and that of the image of *B* have been exactly matched in this way, the reading of the ammeter *A* is taken.

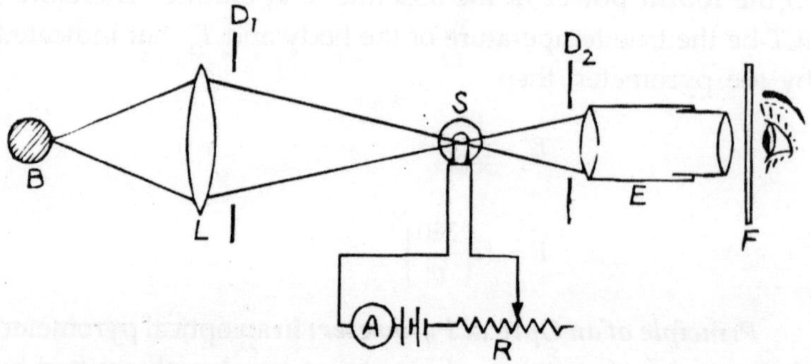

Calculation of Temperature : The brightness of the body *B* is the same as that of the filament *S*. Hence their emissive powers are equal. Thus the temperature of the body is equal to that of the filament which is obtained by

$$i = a + bT + cT^2,$$

where *i* is the current through the filament and *a, b, c* , are constants. These can be determined by calibrating the instrument against known temperatures.

This instrument can measure temperatures from 600 °C to 1500 °C. The range can be raised to 2700 °C by using a rotating sector to cut out a known fraction of the radiation.

Polarising Pyrometer

Another type of optical pyrometer, due to Wanner, is a "polarising pyrometer", which is shown in Fig. The radiation beam starting from the hot body (under test), after passing through a hole falls on a calcite prism *P*. Since calcite is doubly-refracting, it breaks the beam into two components polarised at right angles to each other. These components are indicated by horizontal and vertical arrows in the diagram. They next

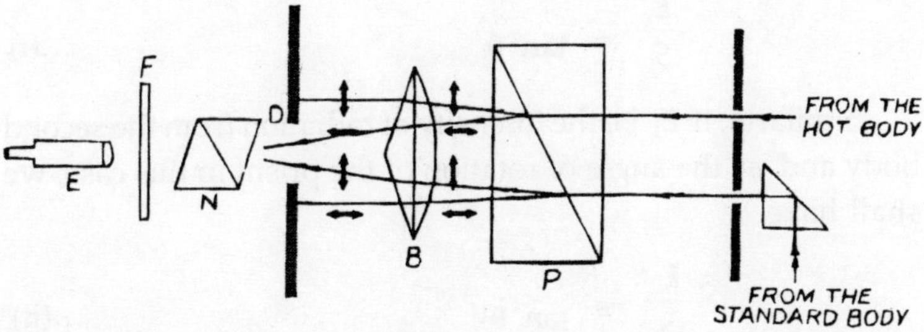

pass through a biprism *B* which bends them in such a way that the horizontal component passes through a diaphragm *D*, while the vertical component is cut off. In the same way, a radiation beam from a standard body at a steady temperature is sent through the system but in this case the vertical component passes, while the horizontal component is cut off. The two mutually perpendicular components emerging from *D* pass through a Nicol prism *N* and then through a red filter *F*. When an observer looks through an eyepiece *E*, two semi-circular patches of light are seen.

Working : The Nicol *N* is rotated. It is found that in one particular position of *N* , the light from the given hot body transmits fully through *N* but the light from the standard body is completely cut off. (The opposite is the case when the Nicol is rotated by 90° from this position). From this position *the Nicol is gradually rotated until the two patches of light appear equally bright.* In this position the light from both the bodies is transmitted equally through N. The angle through which the Nicol has been rotated is carefully measured. The experiment is repeated with a second body at a *known* temperature, the standard body remaining the same.

Calculation of Temperature : Let E_λ be the intensity of radiation (filtered through F) from the hot body under test, and S_λ the intensity of radiation of same colour from the standard body. If θ be the angle through which the Nicol has been rotated, then we can show that

$$\frac{E_\lambda}{S_\lambda} = \tan^2\theta \qquad \text{...(i)}$$

Similarly, if E_λ be the intensity of radiation from the second body and θ' the angle of rotation of the prism in this case, we shall have

$$\frac{E_\lambda'}{S_\lambda} = \tan^2\theta' \qquad \text{...(ii)}$$

From eq. (i) and (ii), we get

$$\frac{E_\lambda}{S_\lambda'} = \frac{\tan^2\theta}{\tan^2\theta'} \qquad \text{...(iii)}$$

Now, from Wien's distribution law (remembering that radiation intensity is proportional to radiation density), we have

$$E_\lambda = A\lambda^{-5}e^{-B/\lambda T}$$

and

$$E_\lambda' = A\lambda^{-5}e^{-B/\lambda T'}$$

where T is the temperature (unknown) of the body under test and T' (known) that of the second body, and A and B are constants.

$$\therefore \quad \frac{E_\lambda}{E_\lambda'} = \frac{B}{e^\lambda}\left(\frac{1}{T'} - \frac{1}{T}\right) \qquad \text{...(iv)}$$

A comparison of eq. (iii) and (iv) gives

$$\frac{\tan^2\theta}{\tan^2\theta'} = \frac{B}{e^\lambda}\left(\frac{1}{T'} - \frac{1}{T}\right)$$

$$\text{or} \quad \log\left(\frac{\tan^2\theta}{\tan^2\theta'}\right) = \frac{B}{\lambda}\left(\frac{1}{T'} - \frac{1}{T}\right)$$

$$\text{or} \quad 2\,(\log\tan\theta - \log\tan\theta') = \frac{B}{\lambda}\left(\frac{1}{T'} - \frac{1}{T}\right)$$

This is of the form

$$\log\tan\theta = a + \frac{b}{T}$$

where a and b are constants and are obtained by calibrating the pyrometer against known temperatures. Thus T may be obtained.

Usually, this pyrometer measures temperatures between 700° and 1400 °C. The range can, however, be raised to 4000 °C by putting absorption glasses in the path of the incident radiation.

Solar Constant

It is defined as the amount of energy received from the sun by the earth per minute per cm² of surface placed normally to the sun's rays at the mean distance of the earth from the sun in the absence of the atmosphere. Its value is 1.937 cal-cm^{-2}-minute^{-1}.

Determination of Solar Constant: The instruments which are employed to measure the solar constant are called "pyrheliometers". Angstrom's pyrheliometer is shown in Fig. It consists of two identical strips of blackened platinum foil, S_1 and S_2. One of them is open to receive radiation from the sun normally; while the other has been protected by a double-walled shield *H*. The junctions of a copper-constantan thermocouple are attached to the backs of S_1 and S_2 . An electric current can be passed through the shielded strip S_2.

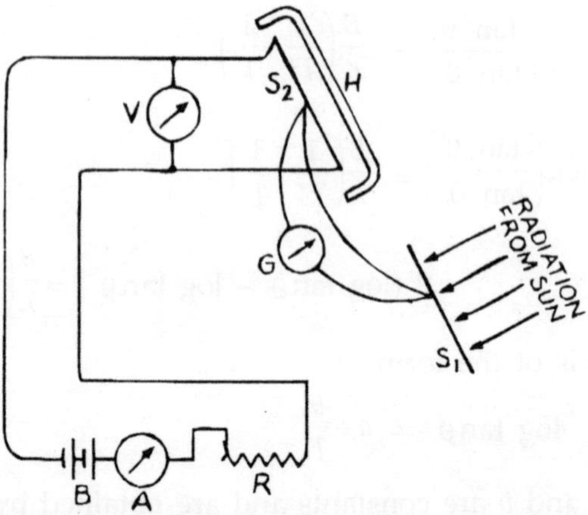

Working : When both S_1 and S_2 are shielded from the radiation, both the junctions are at the same temperature and the galvanometer *G* shows no deflection. When S_1 is allowed to receive radiation from the sun, its temperature rises and the galvanometer is deflected. Now, current is passed through the shielded strip S_2 and is so adjusted that the deflection of the galvanometer reduces to zero. The temperatures of S_1 and S_2 are then the same, and it follows that they must be receiving energy at the same rate. If *i* amp be the current flowing through S_2 and V volt the potential difference across it, then the energy supplied to S_2 is

$$Vi \text{ watt} = Vi \text{ joule/sec}$$

$$= Vi \times 10 \text{ erg/sec}$$

$$= \frac{Vi}{J} \times 10^7 \text{ cal/sec}$$

where *J* is number of ergs/calorie. This is also the energy received per second by the strip S_1 from the sun. If *l* and *b* are the length and breadth of the strip, *a* its absorption coefficient and H the radiation incident on it per second per cm², then the energy absorbed by the strip per second is $Halb$. Thus

$$Halb = \frac{Vi}{J} \times 10^7$$

or $$H = \frac{Vi}{albJ} \times 10^7$$

From this, the intensity of the incident radiation is obtained in calories, from which the solar constant can be calculated.

The experiment is repeated by allowing the strip S_2 to receive the radiation and shielding the strip S_1, and the mean value is obtained. Thus any error due to mechanical inequality in the strips is eliminated.

The atmosphere absorbs a part of the radiation. To eliminate it, the solar constant is found for different elevations of the sun on the same day. If S be the observed value of the solar constant, when the zenith distance of the sun is z, we have

$$S = S_0 a^{\sec z},$$

where S_0 is the true value of the solar constant and a is the transmission coefficient of the atmosphere. The above eq. can be written as

$$\log S = \log S_0 + \sec z \log a.$$

The values of log S are plotted against sec z. A straight line is obtained Fig. Its intercept on the log S-axis gives log S_0. From this, S_0 can be calculated.

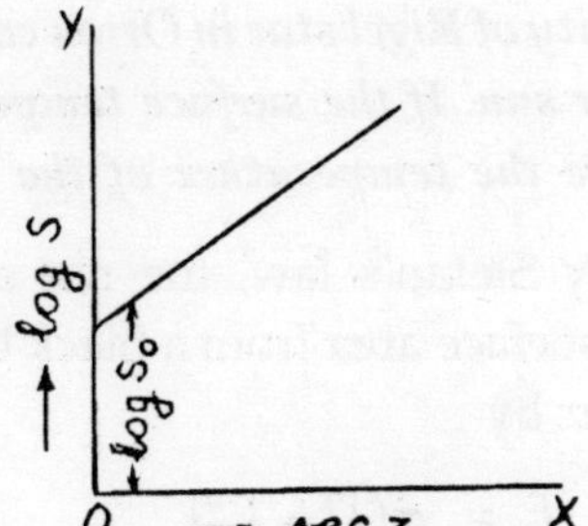

Determination of Temperature of the Sun : The sun consists of a central hot part known as the *"photosphere "*, which is surrounded by a comparatively cooler atmosphere, called "chromosphere." By the temperature of sun we mean temperature of the photosphere.

Let r be the radius and T the absolute temperature of the photosphere of the sun. If we treat it as a black body, then by Stefan's law, the heat radiated by it per unit time is

$$4\pi r^2 = \sigma T^4$$

where σ is the Stelan's constant. If R be the mean distance of the earth from the sun, then this radiated energy will be spread over a sphere of surface area $4\pi R^2$. Hence the energy received per unit time per unit surface area of the earth is

$$\frac{4\pi R^2 \times \sigma T^4}{4\pi R^2}$$

But, this is defined as the solar constant S.

$$\therefore \qquad S = \frac{4\pi r^2 \times \sigma T^4}{4\pi R^2}$$

$$\text{or} \qquad T^4 = \frac{S}{\sigma}\left(\frac{R}{r}\right)^2$$

Thus, knowing the values of S, R, r and σ, the black-body temperature of the sun, T, can be found out.

PROBLEMS

1. ***Luminosity of Rigel star in Orion constellation is 17,000 times that of our sun. If the surface temperature of the sun is 6000 K, calculate the temperature of the star.***

Solution : By Stefan's law, the net energy radiated per second per unit surface area from a black body at temperature T Kelvin is given by

$$E = \sigma(T^4 - T_0^4),$$

where T_0 is the temperature of the surroundings and σ is the Stefan's constant. If $T >> T_0$, we may write

$$E = \sigma T^4.$$

Now, let E_1 and E_2 be the luminosities of the star and the sun respectively , and T_1 and T_2 their temperatures. Then, treating them as black bodies, we get

$$\frac{E_1}{E_2} = \frac{T_1^4}{T_2^4}$$

Here $\quad E_1/E_2 = 17{,}000$

and $\quad T_2 = 6000 \text{ K}.$

$$\therefore \quad 17000 = \frac{T_1^4}{(6000)^4}$$

or $\quad T_1^4 = (6000)^4 \times 17000.$

$$\therefore \quad T_1 = 6000 \times (17000)^{1/4}$$

$$= 6000 \times 11.42$$

$$= 68520 \text{ K}.$$

2. ***At what rate will energy be emitted from a bulb whose filament has a temperature of 3600 K, if at 1800 K the energy is emitted at the rate of 16 watt ?***

Solution : By Stefan's law, the rate of energy emission from a (black) body at Kelvin temperature T is

$$E \propto T^4$$

Thus $\quad \dfrac{E_1}{E_2} = \dfrac{T_1^4}{T_2^4}$

Here $\quad E_1 = 16 \text{ watt},$

$T_1 = 1800 \text{ K},$

$T_2 = 3600 \text{ K}, E_2 = ?$

$$\therefore \quad E_2 = E_1 \times \frac{T_2^4}{T_1^4}$$

$$= 16 \text{ watt} \times \frac{(3600 \text{ K})^4}{(1800 \text{ K})^4}$$

$$= 256 \text{ watt.}$$

3. *A black body radiates heat per unit area at a rate of 10^5 watt at 300 °C. If sun radiates heat per unit area at a rate of 10^9 watt, then compute the temperature of sun.*

Solution : Refer to the last problem.

$$\frac{E_1}{E_2} = \frac{T_1^4}{T_2^4}$$

Here $E_1 = 10^5$ watt,

$T_1 = 300 + 273 = 573$ K

and $E_2 = 10^9$ watt.

$$\therefore \quad T_2^4 = T_1^4 \times \frac{E_2}{E_1}$$

$$= (573)^4 \times \frac{10^9}{10^5} = (5730)^4$$

$$\therefore \quad T_2 = 5730 \text{ K.}$$

4. *A black body is placed in an evacuated enclosure whose walls are blackened and kept at constant temperature of 300 K. Compare the net amount of heat gained or lost by the body when its temperature is (i) 600 K , (ii) 100K.*

Solution: When a black body at Kelvin temperature T is surrounded by another black body at temperature T_0 $(< T)$, the net loss of heat energy per second per unit surface area is given by

$$E_1 = \sigma (T^4 - T_0^4)$$

Here $T = 600$ K

and $T_0 = 300$ K .

$\therefore$ $E_1 = \sigma\,(600^4 - 300^4)$

$= \sigma\,(1215 \times 10^8)$.

If $T_0 > T$, the net gain of energy per second per unit surface area would be given by

$$E_2 = \sigma\,(T_0^{\,4} - T^4).$$

Now $T_0 = 300$ K

and $T = 100$ K.

$\therefore$ $E_2 = \sigma\,(300^4 - 100^4)$

$= \sigma\,(80 \times 10^8)$.

Thus L = 12!5 = 243.

$$\frac{E_1}{E_2} = \frac{1215}{80} = \frac{243}{16}.$$

5. ***Find the amount of energy radiated per minute from a black body at temperature 2000 K if its surface area is 5×10^{-5} meter^2. Stefan's constant is 5.67×10^{-8} joule/(meter^2-second-K^4).***

Solution: According to Stefan's law, the radiant energy (heat) emitted per second per unit surface area of a black body is given by

$$E = \sigma\,(T^4 - T_0^{\,4}),$$

where T is the absolute temperature of the body and T_0 that of the surroundings. If $T >> T_0$, then we can write

$$E = \sigma T^4$$

If A be the surface area of the black body, then the energy emitted by it per minute (= 60 second)

$$= \sigma T^4 \times A \times 60$$

$$= 5.67 \times 10^{-8} \frac{J}{m^2 - s = K^4} \times (2000\ K)^4 \times (5 \times 10^{-5}\ m^2) \times 60\ s$$

$$= 2.72 \times 10^3\ J.$$

6. A black body with an initial temperature of 300 °C is allowed to cool inside an evacuated enclosure surrounded by melting ice at the rate of 0.35 °C/sec . If the mass, specific heat and surface area of the body are 32 gram, 0.10 cal/(gram-°C) and 8 cm² respectively, calculate the Stefan's constant.'

Solution: By Stefan's law, the net energy radiated per cm² from a black body at temperature T Kelvin placed in an enclosure at temperature T_0 is

$$E = \sigma\ (T^4 - T_0^4)\ \text{erg/(cm}^2\text{-sec)},$$

where σ is the Stefan's constant. Therefore, if A cm² be the surface area of the body, the energy (heat) radiated per second from the entire body is given by

$$Q = \sigma\ (T^4 - T_0^4)\ A\ \text{erg/sec}$$

$$= \frac{\sigma\left(T^4 - T_0^4\right)A}{J}\ \text{cal/sec}$$

But Q = mass × sp. heat × fall in temperature per second

$$= ms\ \Delta T\ \text{cal/sec}.$$

$$\therefore \qquad \frac{\sigma\left(T^4 - T_0^4\right)A}{J} = ms\Delta T$$

$$\text{or} \qquad \frac{Jms\Delta T}{\left(T^4 - T_0^4\right)A}$$

Now, $7 = 4.18 \times 10^7$ erg/cal, m = 32 g , s = 0.10cal/(g-K), ΔT =0.35 K/sec, T= 300 + 273 = 573 K, T_0 = 273 K and A = 8 cm².

$$\therefore \qquad \sigma = \frac{(4.18 \times 10^7) \times 32 \times 0.10 \times 0.35)}{[(573)^4 - (273)4] \times 8}$$

$$= \frac{4.18\times10^{7}\times32\times0.10\times0.35}{1.02\times10^{11}\times8}$$

$$= 5.73\times10^{-5} \text{ erg/(cm}^2\text{-sec-K}^4\text{)}.$$

7. A black body of thermal capacity 1000 cal/°C and surface area 500 cm² is kept inside an evacuated enclosure at 27 °C. Find the rate of cooling of the body when its temperature is 127 °C. The Stefan's constant σ = 1.36 × 10⁻¹² cal/(cm²-sec-°C⁴).

Solution : By Stefan's law, the rate of heat loss (energy radiation) from a black body of surface area A is

$$\sigma\ (T^4 - T_0^{\ 4})A,$$

where T and T_0 are Kelvin temperatures of the body and the surroundings respectively. If ΔT be the rate of cooling of the body (mass m, specific heat s), then we have

$$\Delta (T^4 - T_0^{\ 4})A = ms\ \Delta T$$

$$\therefore \quad \Delta T = \frac{\sigma\left(T^4 - T_0^{\ 4}\right)}{ms}$$

Here $T^4 - T_0^{\ 4} = (400)^4 - (300)^4 = 175 \times 10^8$ K⁴ , $A = 500$ cm², $ms = 1000$ cal/°C = 1000 cal/K . (The magnitude of a degree Celsius is same as the magnitude of a Kelvin.)

$$\therefore \quad \Delta T - \left(1.36\times10^{-12}\frac{\text{cal}}{\text{cm}^2-\text{sec}-K^4}\right)\frac{175\times10^8 K^4\left(500cm^2\right)}{1000 \text{ cal/K}}$$

$$= 0.0119 \text{ K/sec}$$

$$= 0.0119 \text{ °C/sec} .$$

8. Calculate the maximum amount of heat which may be lost per second by radiation from a sphere of 10 cm diameter at a temperature of 227 °C when placed in an enclosure at a temperature of 27 °C. Stefan's constant σ = 5.7 × 10⁻¹² watt/ (Cm²-°C⁴).

Solution: The heat loss will be maximum when the body is treated as black body. By Stefan's law, the rate of heat loss from a black body of surface area A is

$$\Delta Q = \sigma (T^4 - T_0^4)A, \quad ...(1)$$

where T and T_0 are Kelvin temperatures of the body and the surroundings respectively.

Here r=227 + 273 = 500 K and T_0 = 27 + 273 = 300 K, so that $T^4 - T_0^4 = (500\text{ K})^4 - (300\text{ K})^4 = 544 \times 10^8\text{ K}^4$, and $A = 4\pi r^2 = 4 \times 3.14 \times (5\text{ cm})^2 = 314\text{ cm}^2$.

Substituting the given values in eq. (i), we get

$$\Delta Q = 5.7 \times 10^{-12} \frac{\text{watt}}{\text{cm}^2 - {}^\circ\text{C}^4} \times (544 \times 10^8\text{ K}^4) \times (314\text{ cm}^2).$$

Now, 1 watt = 10^7 erg/sec .

$$\therefore \Delta Q = \frac{5.7 \times 10^{-12} \times 10^7\text{ erg/sec}}{\text{cm}^2\text{ -K}^4} \times (544 \times 10^8\text{ K}^4) \times (314\text{ cm}^2)$$

$$= 9.736 \times 10^8\text{ erg/sec}$$

$$= \frac{9.736 \times 10^8}{4.18 \times 10^7} = 23.3\text{ cal/sec } [\because 1\text{ cal} = 4.18 \times 10^7\text{ erg}]$$

9. *Deduce the temperature at which a black body loses thermal energy at a rate of 1 watt/cm².*

Given : $\sigma = 5.6 \times 10^{-5}$ erg-cm⁻²-sec⁻¹-deg⁻⁴.

Solution: The energy emitted per second per cm² from a black body is given by

$$E = \sigma T^4.$$

$$\therefore \quad T = \left(\frac{E}{\sigma}\right)^{1/4}$$

Here E = 1 watt/cm² = 10^7erg-sec^{-1} cm². Substituting this and the value of σ, we get

$$T = \left[\frac{10^7 \text{ erg-sec}^{-1}\text{-cm}^{-2}}{5.6\times10^{-5} \text{ erg-cm}^{-2}\text{-sec}^{-1}\text{-deg}^{-4}}\right]^{1/4}$$

$$= 0.65\times10^3 \text{ deg} = 650 \text{ K}.$$

10. A solid copper sphere cools at the rate of 2.8 °C per minute when its temperature is 127 °C. At what rate will a solid copper sphere of twice the radius cool when its temperature is 227 °C, if in both cases, the surroundings are maintained at 27 °C and, the conditions are such that Stefan's law may be applied?

Solution: According to Stefan's law, the rate of cooling of a body (mass m, surface area A , specific heat s) is given by

$$\therefore \quad \Delta T = \frac{\sigma\left(T^4 - T_0^4\right)A}{Jms}$$

For a sphere of radius R and density ρ, we have

$$A = 4\pi R^2$$

and $$m = \frac{4}{3}\pi R^3 \rho .$$

$$\therefore \quad \Delta T = \frac{3\sigma\left(T^4 - T_0^4\right)}{JsR\rho}$$

For the first sphere (T= 400 K, T_0 = 300 K), we have

$$\Delta T = 2.8 \text{ °C/ minute. Thus}$$

$$2.8 = \frac{3\sigma[(400)^4 - (300)^4]}{js(R)\rho} = \frac{3\sigma\left(1.75\times10^{10}\right)}{js(R)\rho} \qquad \ldots(1)$$

For the second sphere of twice radius (T= 500 K and T_0 = 300 K), we shall have

$$\Delta T = \frac{3\sigma[(500)^4 - (300)^4]}{Js(2R)P} = \frac{3\sigma(5.44\times10^{10})}{Js\,(2R)\,\rho} \qquad \ldots(\text{ii})$$

Dividing eq. (ii) by eq. (i), we get

$$\Delta T = 2.8 \times \frac{5.44}{2 \times 1.75} = 4.35°C/\text{minute}.$$

11. The operating temperature of the tungsten filament of a bulb is 3000 K. Its surface area is 0.25 cm² and the emissivity is 0.35. Find the wattage of the bulb, $\sigma = 5.67 \times 10^{-5}$ erg-cm^{-2} sec^{-1} deg^{-4}.

Solution: The energy emitted per second per cm² from the (non-black body) filament at Kelvin temperature T is, by Stefan's law, given by

$$E = e\sigma T^4,$$

where e is the emissivity, and the temperature of the surroundings is ignored. For an area A , the total energy emitted per second (power radiated) is

$$W = e\ \sigma\ T^4 A.$$

Substituting the given values, we get

$$W = 0.35 \times (5.67 \times 10^{-5} \text{ erg-cm}^{-2}\text{- sec}^{-1}\text{- deg}^{-4}) \times (3000 \text{ deg})^4 \times 0.25 \text{ cm}^2$$

$$= 40 \times 10^7 \text{ erg/sec}$$

$$= 40 \text{ watt}.$$

12. Calculate the number of modes of vibration in wave-length range 5000 Å to 5002 Å for radiation in a chamber of volume 100 cm³.

1.17×10^4 joule/min.

13. Calculate the number of modes of vibration in wave-length range Å 5000 to Å 5002 for radiation in a chamber of volume 100 cm³.

Solution: The number of modes, or degrees of freedom, between λ and $\lambda + d\lambda$ is given by

$$df = \frac{8\pi V}{\lambda^4} d\lambda,$$

where V is the volume of the radiation chamber. Substituting the given values, we get

$$df = \frac{8 \times 3.14 \times 100\text{cm}^3 \times (2 \times 10^{-8}\text{cm})}{(5000 \times 10^{-8}\ \text{cm})^4}$$

$$= 8.0 \times 10^{12}.$$

14. ***Calculate the average energy of Planck's oscillator of frequency 0.60 × 10^{14} sec^{-1} at 1800 K. (h = 6.63 × 10^{-34} J–s, k = 1.38 × 10^{-23} J/K.) If the oscillator be classical, then what will be its average energy?***

Solution: The average energy of Planck's oscillator is given by

$$\bar{E}_{Planck} = \frac{h\nu}{e^{h\nu/kT} - 1}$$

Here $h\nu = (6.63 \times 10^{-34}\ \text{Js})\ (0.60 \times 10^{14}\ \text{s}^{-1})$

$$= 3.98 \times 10^{-20}\ \text{J, so that}$$

$$\frac{h\nu}{kT} = \frac{3.98 \times 10^{-20}\ \text{J}}{(1.38 \times 10^{-23}\ \text{J/K}) \times 1800\text{K}} = 1.60$$

and $e^{h\nu/kT} = e^{1.60} = (2.718)^{1.60} = 4.95$

$$\therefore \quad \bar{E}_{Planck} = \frac{3.98 \times 10^{-20}\ \text{J}}{4.95 - 1} = 1.01 \times 10^{-20}\ \text{J}$$

The average energy of classical oscillator is

$$\bar{E}_{clasical} = kT$$

$$= (1.38 \times 10^{-23}\ \text{J/K})\ (1800\ \text{K})$$

$$= 2.48 \times 10^{-20}\ \text{J}.$$

15. Earth receives 1.90 calories of heat per cm² per minute. Calculate the surface temperature of the sun. Given : distance of earth from sun = 1.50 × 10⁸ km, diameter of sun= 1.39 × 10⁶ km, Stefan's constant = 1.37 × 10⁻¹² cal/(cm²-sec-deg⁴).

Solution: If r be the radius of the sun, R the mean distance of the earth from the sun, S the solar constant and σ the Stefan's constant, then the temperature T of the sun is given by

$$T^4 = \frac{S}{\sigma}\left(\frac{R}{r}\right)^2$$

The solar constant S is energy received by the earth from the sun per cm² per minute. Thus

$$\frac{S}{\sigma} = \frac{1.90\,\text{cal}/\left(\text{cm}^2-\text{min}\right)}{1.37\times10^{-12}\,\text{cal}/\left(\text{cm}^2-\text{sec}-\text{K}^4\right)}$$

$$= \frac{\frac{1.90}{60}\text{cal}/\left(\text{cm}^2-\text{sec}\right)}{1.37\times10^{-12}\,\text{cal}/\left(\text{cm}^2-\text{sec}-\text{K}^4\right)}$$

$$= 2.31 \times 10^{10}\ \text{K}^4$$

and $$\frac{R}{r} = \frac{1.50\times10^8\,\text{km}}{0.695\times10^6\,\text{km}} = 215.8$$

$$\therefore \quad T^4 = \frac{S}{\sigma}\left(\frac{R}{r}\right)^2$$

$$= 2.31\times10^{10}\text{K}^4\times(215.8)^2$$

$$= 1076\times10^{12}\text{K}^4$$

$$\therefore \quad T = 5.727\times10^3\text{K} = 5727\text{K}$$

This is the *brightness* or *black-body* temperature of the sun's surface.

16. Energy falling on 1.0 m² area placed at right angles to a sunbeam just out-side the earth's atmosphere is 1.35 kJ in one second. Find sun's surface temperature. Mean distance of earth from sun is 1.50 × 10⁸ km, mean diameter of sun = 1.39 x 10⁶ km and Stefan's constant is 5.67 × 10⁻⁸ watt-m⁻²-K⁻⁴.

Solution: If r be the radius of the sun, R the mean distance of the earth from the sun, S the solar constant and σ the Stefan's constant, then the temperature T of the sun's surface is given by

$$T^4 = \frac{S}{\sigma}\left(\frac{R}{r}\right)^2$$

Now
$$\frac{S}{\sigma} = \frac{1.35\,\text{kilo}-\text{joule}/\text{meter}^2-\text{sec}}{5.67\times10^{-8}\,\text{watt}/\text{meter}^2-\text{K}^4}$$

$$= \frac{1.35\times10^{3}\,\text{watt}/\text{meter}^2}{5.67\times10^{-8}\,\text{watt}/\text{meter}^2-\text{K}^4} \qquad \left[\because \text{watt} = \frac{\text{joule}}{\text{sec}}\right]$$

$$= 2.38 \times 10^{10}\ \text{K}^4$$

and
$$\frac{R}{r} = \frac{1.50\times10^{8}\,\text{km}}{0.695\times10^{6}\,\text{km}} = 215.8$$

$$\therefore \quad T^4 = \left(2.38\times10^{10}\,\text{K}^4\right)(215.8)^2$$

$$= 1108 \times 10^{12}\ \text{K}^4$$

$$\therefore \quad T = 5.770 \times 10^3\ \text{K}$$

$$= 5770\ \text{K}$$

17. The angular diameter of the sun is 32′ and it is treated as a black body. Calculate its surface temperature. Solar constant S is 1.34 kilowatt/meter² and Stefan's constant σ is 5.67 × 10⁻⁸ watt/(meter²-K⁴).

Solution: The Kelvin temperature T of the sun is given by

$$T^4 = \frac{S}{\sigma}\left(\frac{R}{r}\right)^2 \quad \text{...(i)}$$

Now, $S = 1.34 \times 10^3$ watt/meter^2

and $\sigma = 5.67 \times 10^{-8}$ watt/(meter^2-K^4).

$$\therefore \quad \frac{S}{\sigma} = \frac{1.34 \times 10^3}{5.67 \times 10^{-8}} = 2.36 \times 10^{10}\,\text{K}^4$$

Further, the angular diameter is 32' *i.e.* the angular radius of the sun, r/R , is 16'. Thus

$$\frac{r}{R} = 16' = \left(\frac{16}{60}\right)^{\circ} = \frac{16}{60} \times \frac{3.14}{180} \text{ radian}$$

$$\therefore \quad \frac{R}{r} = \frac{60 \times 180}{16 \times 3.14} = 125$$

Substituting the values of $\frac{S}{\sigma}$ and $\frac{R}{r}$ in eq. (i), we get

$$T^4 = (2.36 \times 10^{10}\,\text{K}^4)(215)^2$$

$$= 1091 \times 10^{12}\,\text{K}^4$$

$$\therefore \quad T = 5.747 \times 10^3\,\text{K} = 5747\,\text{K}.$$

18. ***Show that the pressure exerted by sunlight is equal to 4.66 × 10^{-6} N/m^2. Given : solar constant = 2 × 10^4 cal/(min-m^2) and J = 4.2 joule/cal.***

Solution: The solar constant S is the solar energy reaching the earth per unit time per unit surface area placed normally to the sunlight.

$$S = 2 \times 10^4 \text{cal/(min-m}^2)$$

$$= \frac{2 \times 10^4}{60} \text{cal/}\left(\text{s-m}^2\right)$$

$$= \frac{2 \times 10^4}{60} \times 4.2\,\text{J}/(\text{s-m}^2)$$

$$= 1.4 \times 10^3\,\text{J}/(\text{s-m}^2).$$

If u be the energy density (energy per unit volume), then the amount of solar energy reaching per second per unit surface area of the earth

$$= cu,$$

which is the amount of energy contained in a cylinder of length c and cross-sectional area unity. This is same as the solar constant S. Thus

$$cu = S = 1.4 \times 10^3\,\text{J}/(\text{s-m}^2).$$

$$\therefore \quad u = \frac{s}{c} = \frac{1.4 \times 10^3\,\text{J}(\text{s-m}^2)}{3 \times 10^8\,\text{m/s}}$$

$$= 4.66 \times 10^{-6}\,\text{J/m}^3.$$

We know that the surface absorbing the radiation falling upon it experiences a pressure p, given by

$$p = u = 4.66 \times 10^{-6}\,\text{N/m}^2.$$

19. The radiation energy received normally at the earth surface from the sun is 1.4 × 10³ joule per meter² per second. Let earth's surface by taken as a perfect absorber and its diameter be 6 × 10⁷ meter. Calculate the radiation pressure and the total force at the earth's surface due to solar radiation.

Solution: The pressure p experienced by earth's surface due to solar radiation perfectly absorbed by it equals the energy density u. Thus

$$p = u = \frac{S}{c}$$

where S is solar constant (energy received normally per unit time per unit surface area) and c is speed of light.

$$\therefore \qquad p = \frac{1.4\times10^3\ \text{J}/(\text{m}^2\text{-s})}{3\times10^8\ \text{m/s}}$$

$$= 4.66 \times 10^{-6}\ \text{J/m}^3$$

$$= 4.66 \times 10^{-6}\ \text{N/m}^2$$

Total force due to solar radiation on earth is pressure x surface area. Thus

$$F = p\times 4\pi r^2$$

$$= 4.66 \times 10^{-6} \times 4 \times 3.14 \times (3 \times 10^7)^2$$

$$= 5.27 \times 10^{10}\text{N}.$$

20. ***Find the quantity of energy radiated from 1 cm² of a surface in one second by a black body if the maximum energy density corresponds to a wave-length of 4840 Å. (Wien's constant b = 2.9 × 10⁻³m-K and Stefan's constant σ = 5.67 × 10⁻⁸ Wm⁻² K⁻⁴.***

Solution: Let T be the absolute temperature of the body. By Wien's law, we have

$$T = \frac{b}{\lambda_m} = \frac{2.9\times10^{-3}\ \text{m-k}}{4840\times10^{-10}\ \text{m}} = 5992\,\text{K}$$

By Stefan's law, the rate of energy emission per unit area of the surface is

$$E = \sigma T^4$$

$$= 5.67 \times 10^{-8}\ \text{W}/(\text{m}^2\text{-K}^4) \times (5992\ \text{K})^4$$

$$= 5.67 \times 10^{-8}\ \text{J}/(\text{m}^2\text{-s-K}^4) \times (5992\ \text{K})^4$$

$$= 7.31 \times 10^7\ \text{J}/(\text{m}^2\text{-s})$$

$$= 7.31 \times 10^3\ \text{J}/(\text{cm}^2\text{-s})$$

21. ***Using Wien' displacement law, estimate the temperature of the sun.***

Given λ_m = 4900 Å and Wien's constant = 0.292 cm-K.

Solution: For a black body at Kelvin temperature T, the Wien's displacement law states that

$$\lambda_m T = b$$

where λ_m is the wave-length at which the spectral radiancy of the body is maximum, and b is Wien's constant.

For the sun it is given that $\lambda_m = 4900$ Å $= 4900 \times 10^{-8}$ cm and $b = 0.292$ cm-K. Hence the temperature of the sun is

$$T = \frac{b}{\lambda_m} = \frac{0.292 \text{ cm-K}}{4900 \times 10^{-8} \text{ cm}}$$

$$= 5.959 \times 10^3$$

$$= 5959 \text{ K}.$$

22. ***From Wien's law, we have***

$$\lambda_m T = \frac{hc}{4.965\,k} = b,$$

the letters having their usual significance. Calculate the constant b. Abbot's measurements show that λ_m for the solar radiation is 4753 Å. Calculate the temperature of the surface of the photosphere of the sun. Given h = 6.63 x 10^{-34} joule-sec, k = 1.38 x 10^{-23} joule/K and c = 3.0 x 10^8 meter/sec.

Solution: $b = \dfrac{hc}{4.965\,k}$

$$= \frac{(6.63 \times 10^{-34} \text{ joule-sec})(3.0 \times 10^8 \text{ meter/sec})}{4.965 \times 1.38 \times 10^{-23} \text{ joule/K}}$$

$$= 2.90 \times 10^{-3} \text{ meter-K}$$

For the solar radiation, $\lambda_m = 4753$ Å. Therefore, from Wien's law $\lambda_m T = b$, the temperature of sun is given by

$$T = \frac{b}{\lambda_m} = \frac{2.90 \times 10^{-3} \text{ meter-K}}{4753 \times 10^{-10} \text{ meter}}$$

$$= 6.10 \times 10^3 = 6100 \text{ K}.$$

This is the *colour* temperature of the sun. It is higher than the *brightness* temperature.

23. Wien's displacement law states : $\lambda_m T = \frac{hc}{5k} = b$ and it is given that $h = 6.6 \times 10^{-34}$ J-s, $c = 3 \times 10^8$ m/s, $k = 1.38 \times 10^{-23}$ J/K. Find the value of b. If λ_m for the moon is 14 microns, find its temperature.

Hint: 1 micron = 10^{-6} meter.

2.87×10^{-3} meter-K, 205 K.

24. Given : Wien's constant b = 0.3 cm-K, estimate the temperatures at which a body would appear red and blue. The corresponding wave-lengths of maximum emission, $\lambda_m \approx$ 7500 Å and 5000 Å respectively.

Solution: By Wien's law;

$$\lambda_m T = b$$

Thus for red appearance, the required temperature is given by

$$T = \frac{b}{\lambda_m} = \frac{0.3\,\text{cm-K}}{7500 \times 10^{-8}} = 4000\text{K}$$

Similarly, for blue appearance, the temperature is

$$T = \frac{0.3}{5000 \times 10^{-8}} = 6000\text{K}$$

25. What is the wave-length of the maximum intensity radiation, radiated from a source having temperature 3000 K. The Wien's constant is 0.3 x 10^{-2} m-K.

Solution: By Wien's law

$$\lambda_m T = b$$

$$\therefore \qquad \lambda_m = \frac{b}{T} = \frac{0.3\times10^{-2}\,\text{m-K}}{3000\,\text{K}}$$

$$= 0.1 \times 10^{-5}\ \text{m}$$

$$= (0.1 \times 10^{-5}) \times 10^{10}$$

$$= 10000\ \text{Å}.$$

26. *Compute the frequency corresponding to the maximum energy density in the radiation emitted from a black body at temperature 1000 K. K = 1.38 × 10⁻²³ joule/K, h = 6.6 × 10⁻³⁴ joule-sec.*

Solution: If λ_m be the wave-length corresponding to maximum energy in black-body radiation at Kelvin temperature T, then we have, by Wien's law

$$\lambda_m T = \frac{hc}{4.965k} \simeq \frac{hc}{5k}$$

The corresponding frequency is

$$\nu = \frac{c}{\lambda_m} = \frac{5kT}{h}$$

$$= \frac{5\times\left(1.38\times10^{-23}\,\text{joule/K}\right)\times1000\,\text{K}}{6.6\times10^{-34}\,\text{joule/sec}}$$

$$= 1.045 \times 10^{14}\ \text{per sec.}$$

27. *A body at 1500 K emits maximum energy at a wave-length 20,000 Å. If the sun emits maximum energy at wave-length 5500 Å, what would be the temperature of the sun ?*

Solution: By Wien's displacement law

$$\lambda_m T = \text{constant}$$

or

$$\lambda_m T = \lambda_m' T'$$

or $$T' = \frac{\lambda_m T}{\lambda_m{'}}$$

$$= \frac{20{,}000\,\text{Å} \times 1500\,\text{K}}{5500\,\text{Å}}$$

$$= 5454 \text{ K.}$$

28. The black-body spectral energy distribution of radiation from moon shows two wave-length maxima at* 14μ *and 5000 Å. What is the significance of such an observation ? Calculate the corresponding temperature. Given : b = 0.3 cm-deg .

Solution. The wave-length maximum at 14 u is due to moon's own radiation, while that at 5000 Å coincides with the maximum in solar radiation because solar radiation is reflected from moon's disc. According to Wien's displacement law,

$$\lambda_m T = b$$

For moon $$T = \frac{b}{\lambda_m} = \frac{0.3\,\text{cm-K}}{14 \times 10^{-4}\,\text{cm}} \qquad [\because 1\mu = 10^{-4}\,\text{cm}]$$

$$= 214 \text{ K}$$

For sun : $$T = \frac{b}{\lambda_m} = \frac{0.3\,\text{cm-K}}{5000 \times 10^{-8}\,\text{cm}} = 6000\,\text{K}$$

4

Distribution of Energy

Condition for Equilibrium : Let us consider two macroscopic system A and A' having energies denoted by E and E' respectively. Let $\Omega(E)$ be the number of microstates accessible to A when its energy lies between E and $E + \delta E$, and $\Omega'(E')$ the number of microstates accessible to A' when its energy lies between E' and $E' + \delta E'$. To an excellent approximation, we can assume all the $\Omega(E)$ states of A lying in the energy-interval between E and $E + \delta E$ simply to have an energy equal to E'. The same is true for A$'$.

Let us assume that the systems A and A' are in thermal contact, that is, they, are free to exchange energy (in the form of heat). Although the energy of each system separately is then not constant, but the combined system A* = A + A$'$ is isolated so that its total energy E^* must remain unchanged. Therefore, neglecting any interaction energy, we can write

$$E + E' = E^* = \text{constant}. \qquad \text{...(i)}$$

Let us now consider the situation when A and A $'$ have attained thermal equilibrium, that is, when the combined *isolated* system A* is in equilibrium. A* is equally likely to be found in each one of its accessible states (basic postulate of statistics). Let Ω^*_{total} be the total number of states accessible to A*, and $\Omega^*(E)$ the number of states of A* which are such that the sub-system A has an energy equal to *E*. Then, the probability *P(E)* that A has energy equal to *E (i.e.* it lies in the interval between *E* and $E+\delta E$), is given by

$$P(E) = \frac{\Omega^*(E)}{\Omega^*_{total}} = C\Omega^*(E), \qquad \text{...(ii)}$$

where $C\ (=1/\Omega^*_{total})$ is some constant independent of *E*.

The number of states $\Omega^*(E)$ of A* can be expressed in terms of the number of states of A and A′. When A has energy *E*, it can be in any one of its $\Omega(E)$ possible states. Then, by e.q. (i), the system A′ must have energy $E'(=E^*-E)$, and it can be in any one of its $\Omega'(E') = \Omega'(E^*-E)$ possible states. Since every state of A can be combined with every state of A′ to give a different possible state of the combined system A*, the number of distinct states of A* when A has an energy *E* is given by the product

$$\Omega^*(E) = \Omega(E)\,\Omega'(E^*-E).$$

Substituting this in eq. (ii), the probability that the system A has an energy *E* is given by

$$P(E) = C\,\Omega(E)\,\Omega'(E^*-E). \qquad \text{...(iii)}$$

Now, according to the basic principle of statistical mechanics, the equilibrium state is one in which *E*, and correspondingly *(E* -E)* or *E′*, have values such that the probability *P(E)* is maximum. To investigate the condition for *P(E)* to be maximum, it is more convenient to investigate the condition for $\log_e P(E)$ to be maximum . The eq. (iii) can be written as

$$\log_e P(E) = \log_e C + \log_e \Omega(E) + \log_e \Omega'(E'), \quad \text{...(iv)}$$

where $E' = E^* - E$. The value of E which corresponds to the maximum of $\log_e P(E)$ is determined by the condition

$$\frac{\partial\{\log_e P(E)\}}{\partial E} = 0 \quad \text{...(v)}$$

Using eq. (iv), the condition (v) becomes

$$\frac{\partial}{\partial E}\log_e \Omega(E) + \frac{\partial}{\partial E}\log_e \Omega'(E') = 0$$

or

$$\frac{\partial}{\partial E}\log_e \Omega(E) + \frac{\partial}{\partial E'}\log_e \Omega'(E')\frac{\partial E'}{\partial E} = 0$$

But $\frac{\partial E'}{\partial E} = -1$ (because A and A′ *exchange* energy).

$$\therefore \boxed{\frac{\partial}{\partial E}\log_e \Omega(E) = \frac{\partial}{\partial E'}\log_e \Omega'(E').} \quad \text{...(vi)}$$

Thus, for equilibrium of the two systems, the function $\frac{\partial}{\partial E}\log_e \Omega(E)$ must remain constant. This function is denoted by βE. Thus

$$\beta(E) = \frac{\partial}{\partial E}\log_e \Omega(E) = \frac{1}{\Omega(E)}\frac{\partial \Omega(E)}{\partial E}.$$

and

$$\beta'(E') = \frac{\partial}{\partial E'}\log_e \Omega'(E') = \frac{1}{\Omega'(E')}\frac{\partial \Omega'(E')}{\partial E'}.$$

The condition for equilibrium is therefore

$$\boxed{\beta(E) = \beta'(E').} \quad \text{...(vii)}$$

The relation (vii) is the fundamental condition which determines the particular value of the energy of A (and the corresponding value of the energy of A') which occurs with the greatest probability $P(E)$.

Identification of β with 1/kT : According to the definition $\beta(E) = \frac{\partial}{\partial E}\log_e \Omega(E) = \frac{1}{\Omega(E)}\frac{\partial \Omega(E)}{\partial E}$, the parameter β has the dimensions of reciprocal of energy. Let us express $1/\beta$ as a multiple of some positive constant k having dimensions of energy :

$$\boxed{\frac{1}{\beta} = kT.}$$

The quantity T thus defined provides a measure of energy in units of the quantity k.

Now, the condition (vii) for thermal equilibrium is satisfied if

$$\boxed{T = T'.}$$

Since equality of parameter T for the two systems is a condition for thermal equilibrium, the parameter T may be identified with temperature. (We know that two systems in thermal equilibrium are at the same temperature). Infact, T is called the *absolute temperature* of the system and is expressed in degrees. The constant k then has dimensions of energy/ degree and is now known as Boltzmann's constant.

The eq. (vii) can now be written as

$$\beta(E) = \beta'(E') = \frac{1}{kT} = \frac{1}{\Omega(E)}\frac{\partial \Omega(E)}{\partial E}.$$

Thermodynamic Probability

For a given macrostate of a system in thermal equilibrium, there may be a large number of accessible microstates. *The total number of microstates corresponding to a given macrostate is called the thermodynamic probability of that particular macrostate.*

Let us consider a system containing a large number N of

molecules. A macrostate of the system can be expressed in terms of the distribution of the N molecules among various cells in the phase space. Suppose, in a particular macro- state, there are $n_1, n_2, n_3, \ldots, n_r$ molecules in the first, second, third, ...r th cell respectively. The number of ways this can be done is

$$\frac{N!}{n_1!\, n_2!\, n_3! \ldots\ldots n_r!}.$$

Each different way corresponds to a different microstate. Thus

$$\frac{N!}{n_1!\, n_2!\, n_3! \ldots\ldots n_r!}.$$

is the thermodynamic probability of the particular macrostate.

If m molecules (say) go from the i th cell to the j th cell and m molecules come from the j th cell to the i th cell, then the macrostate of the system remains the same, but the microstate changes.

Entropy and Probability : Boltzmann's Relation : We know that a given system so alters itself that in the state of thermal equilibrium its entropy is maximum. On the other hand, statistically, the system alters in the direction of increasing probability and the equilibrium state of the system is the most probable state *i.e.* the state of maximum thermodynamic probability. Thus, in equilibrium state, both the entropy and the thermodynamic probability are maximum. This led Boltzmann to conclude that *the entropy of a system is a function of the probability of the state of the system.* That is,

$$S = f(\Omega), \qquad \ldots(i)$$

where S is the entropy and Ω is the thermodynamic probability of the state of the system (or the total number of microstates corresponding to the given macrostate of the system).

To find out the nature of the function, let us consider two independent systems having entropies S_1 and S_2, and probabilities Ω_1 and Ω_2. Then

$$S_1 = f(\Omega_1) \quad \text{and} \quad S_2 = f(\Omega_2).$$

The entropy of the combined system is

$$S = S_1 + S_2 \qquad \text{(entropy is additive)}$$

$$= f(\Omega_1) + f(\Omega_2)\,. \qquad \text{...(ii)}$$

But the probability, being multiplicative, for the combined system is $\Omega_1\,\Omega_2$. Hence, by eq. (i), the entropy of the combined system should be

$$S = f(\Omega_1\,\Omega_2). \qquad \text{...(iii)}$$

Comparing eq. (ii) and (iii), we get

$$f(\Omega_1\,\Omega_2) = f(\Omega_1) + f(\Omega_2)\,. \qquad \text{...(iv)}$$

The solution of eq. (iv) determines f. To solve it, we first differentiate it partially with respect to Ω_1, and then with respect to Ω_2. This gives

$$\Omega_2 f'(\Omega_1\,\Omega_2) = f'(\Omega_1) \qquad \left[f' \text{ is } \frac{\partial f}{\partial \Omega}\right]$$

and

$$\Omega_1 f'(\Omega_1\,\Omega_2) = f'(\Omega_2)$$

Dividing:

$$\frac{\Omega_2}{\Omega_1} = \frac{f'(\Omega_1)}{f'(\Omega_2)}$$

or

$$\Omega_1 f'(\Omega_1) = \Omega_2 f'(\Omega_2)$$

Generalising this relation, we can write

$$\Omega f'(\Omega) = \text{constant} = k \text{ (say)}$$

or $$f'(\Omega) = \frac{k}{\Omega}.$$

Integration gives

$$f(\Omega) = k \log_e \Omega + C,$$

where C is the integration constant. Substituting this result in eq. (i), we get

$$S = k \log_e \Omega + C. \qquad \text{...(v)}$$

Now, according to Nernst heat theorem, the entropy of a thermodynamic system tends to zero as its temperature tends to absolute zero (as $T \to 0$; $S \to 0$). Further, near absolute zero, the thermodynamic probability $\Omega \to 1$. Thus

$$0 = k \log_e 1 + C$$

or $$C = 0.$$

$$\therefore \quad \boxed{S = k \log_e \Omega.}$$

This is the Boltzmann's entropy-probability relation. The constant k appearing in the relation has been identified as the Boltzmann's universal constant.

Ideal Gas Equation : Suppose, a vessel of volume V_1 contains 1 mole of an ideal gas. The number of molecules in the gas is N (Avogadro number). The probability of a molecule of being found in a small part of volume V of the vessel is V/V_1. The probability of any other molecule of being found in the same small part is also V/V_1. Therefore, the probability of both the molecules of being found simultaneously in the same part is $(V/V_1)^2$, because probability is multiplicative. Hence the probability of all the N molecules of being found in the volume V is

$$\Omega = \left(\frac{V}{V_1}\right)^N.$$

Putting this value of Ω in Boltzmann's entropy-probability relation $S = k \log_e \Omega$, we have

$$S = k \log_e \left(\frac{V}{V_1}\right)^N$$

or

$$S = kN(\log_e V - \log_e V_1).$$

Differentiating it with respect to V (remember that V_1 is constant), we have

$$\frac{\partial S}{\partial V} = kN\frac{1}{V} \qquad \text{...(i)}$$

The first and second laws of thermodynamics, $dQ = dU + dW = dU + p\,dV$ and $dQ = T\,dS$, give

$$dU + p\,dV = T\,dS.$$

Differentiating it with respect to V at constant temperature, we have

$$\left(\frac{\partial U}{\partial V}\right)_T + p = T\left(\frac{\partial S}{\partial V}\right)_T.$$

But, for an ideal gas, $\left(\frac{\partial U}{\partial V}\right)_T = 0$ (Joule's law).

$$\therefore \left(\frac{\partial S}{\partial V}\right)_T = \frac{p}{T} \qquad \text{...(ii)}$$

Comparing eq. (i) and (ii), we get

$$\frac{kN}{V} = \frac{p}{T}.$$

or

$$pV = kNT.$$

But $k N = R$ (universal gas constant).

$$\therefore \quad pV = RT.$$

This is ideal gas equation.

Change in Entropy of an Ideal Gas in Isothermal Expansion: Suppose, a vessel of volume V_f has μ mole of an ideal gas. The number of molecules in the gas is μN, where N is Avogadro's number.

Suppose the vessel is divided by an imaginary wall and its volume on one side of the wall is V_i. The probability of all the μN molecules of being found in the volume V_i is

$$\Omega_i = \left(\frac{V_i}{V_f}\right)^{\mu N}.$$

If the wall be removed, then the probability of all the molecules of being found in the whole volume V_f of the vessel will be 1, that is, $\Omega_f = 1$. Therefore,

$$\frac{\Omega_i}{\Omega_f} = \left(\frac{V_i}{V_f}\right)^{\mu N}. \quad \text{...(i)}$$

Suppose, when all the molecules are in the volume V_i then the entropy of the gas is S_i; and when all the molecules occupy the whole volume V_f, the entropy becomes S_f. Then, by the entropy-probability relation $S = k \log_e \Omega$, we have

$$S_f - S_i = k \log_e \Omega_f - k \log_e \Omega_i = k \log_e \left(\frac{\Omega_f}{\Omega_i}\right).$$

But $\dfrac{\Omega_f}{\Omega_i} = \left(\dfrac{V_f}{V_i}\right)^{\mu N}$, by eq. (i).

$$\therefore\ S_f - S_i = k \log_e \left(\frac{V_f}{V_i}\right)^{\mu N}$$

$$= k\mu N \log_e \left(\frac{V_f}{V_i}\right)$$

$$= \mu R \log_e \left(\frac{V_f}{V_i}\right). \qquad [\therefore\ kN = R]$$

***Statistical Interpretation of the Second Law of Thermodynamics*:** The second law of thermodynamics is concerned with the direction in which any natural process occurs in a system. According to the entropy-interpretation of the law *"any natural process in a system occurs in such a direction that results in an increase in the entropy of the system"*. (In the equilibrium state the entropy remains unchanged). Statistically, an isolated system tends to approach a state (macrostate) of large probability where the number of microstates accessible to it is larger than initially. (If the system is already in its most probable state, it remains in equilibrium). This means that the direction of natural processes is controlled by the laws of probability. Hence, statistically, the second law should be modified as :

A natural process occurring in a system has a large probability of producing a net increase in the entropy of the system and the surroundings.

The probability of a decrease in entropy is very small, but not zero. Under very unusual conditions, the entropy may decrease. Such events are known as 'fluctuations'. Some examples to demonstrate the statistical nature of the second law of thermodynamics are the following :

(i) Let us consider the conduction of heat from a hot stove to a kettle of water. According to the statistical interpretation of the second law, we assert, *not* that heat *must* flow from the stove to the water, but that it is highly probable that heat will flow from the stove to the water. It is to be understood that there is always a chance, although an extremely small one, that the reverse may take place.

(ii) Suppose that a vessel containing a gas is divided into two equal compartments by a partition in which there is a small trap door. On the average, the molecules of the gas will be equally divided between the two compartments, but occasionally there may be more molecules in one compartment than in the other. Similarly, there are chances, although extremely small, when all the molecules are confined to one compartment only. According to the statistical interpretation of the second law, we would say that there is a small probability of all the molecules being confined to one compartment, but a much larger probability of an even distribution of molecules between the two compartments.

The second law tells not only the possible events, it tells which events are most probable. It does not say that the process which results in an increase of order is impossible; but it says that the processes which increase order are much less probable than the processes which increase disorder.

System in Contact with a Heat Reservoir* : *Boltzmann Canonical Distribution : In practice, we come across systems which are *not* isolated, but exchange heat with their large environment. To study such systems, we consider a small system A' in thermal contact with a large heat reservoir A′ at absolute temperature T. A' is so large compared to A that its temperature does not change by small exchange of energy with A . Then, at equilibrium, the probability P_i of finding the system A in any one particular state i of energy E_i , is

$$P_i = C e^{-\beta E_i},$$

where $\beta = 1/kT$ and C is constant of proportionality, independent of i. This probability distribution is known as 'Boltzmann canonical distribution.' Let us prove it.

Let $\Omega'(E')$ be the number of states accessible to A′ when its energy is equal to E' (*i.e.*, when it lies between E' and $E' + \delta E$, where δE is very small compared to the separation between the energy states of A , but large enough to contain many possible states of A′). Although the reservoir A′ can have any energy E', but the combined system A* ($\equiv$ A +A′), being isolated, should have a constant total energy E^* (say). Therefore, by energy conservation, when A′ is in its state i of energy E_i, the reservoir A′ must have an energy

$$E' = E^* - E_i . \qquad ...(i)$$

But, when A is in this *one* definite state i, the number of states accessible to the combined system A* is simply the number of states $\Omega'(E^* - E_i)$ accessible to A′. Now, the *isolated* system A* is equally likely to be found in each one of its accessible states (basic statistical postulate). Therefore, the probability Pi of A being in the state i is proportional to the number of states accessible to A* (when A is known to be in the state i). That is

$$P_i \propto \Omega'(E^* - E_i). \qquad ...(ii)$$

Since A is very much small than A′ (reservoir), we can write

$$E_i << E^*$$

so that

$$E' \simeq E^* \qquad \text{[by eq (i)]}$$

and we can expand $\log_e \Omega'(E^* - E_i)$ by Taylor's expansion as

$$\log_e \Omega'(E^* - E_i) = \log_e \Omega'(E^*) - E_i \left[\frac{\partial(\log_e \Omega')}{\partial E'}\right]_{E'=E^*},$$

(neglecting higher terms)

or
$$\log_e \Omega'(E^* - E_i) = \log_e \Omega'(E^*) - \beta E_i, \qquad ...(iii)$$

where $\beta = \left[\frac{\partial(\log_e \Omega')}{\partial E'}\right]_{E'=E^*}$, which has been identified with $1/kT$. Since the derivative $\partial(\log_e \Omega')\partial E'$ has been evaluated at the fixed energy $E' = E^*, \beta\ (= 1/kT)$ is simply the *constant* temperature parameter of the heat reservoir A′. Hence eq. (iii) gives

$$\Omega'(E^* - E_i) = \Omega'(E^*)e^{-\beta E_i}.$$

Making this substitution in eq. (ii), we get

$$P_i \propto \Omega'(E^*)e^{-\beta E_i}$$

Since $\Omega'(E^*)$ is a constant, independent of i, the last expression may be written as

$$\boxed{P_i = Ce^{-\beta E_i}.} \quad ...(iv)$$

where C is a constant independent of i. This probability distribution is 'Boltzmann canonical distribution'. The exponential factor $e^{-\beta E_i}$ is called the 'Boltzmann factor'.

The constant C in eq. (iv) can be determined by the (normalization) condition that the total probability of *all* possible states of the system A must be unity, that is,

$$\sum_i P_i = 1$$

Now eq. (iv) gives

$$\sum_i P_i = C\sum_i e^{-\beta E_i} = 1$$

or

$$C = \frac{1}{\sum_i e^{-\beta E_i}}$$

Making this substitution in eq. (iv), we get

$$P_i = \frac{e^{-\beta E_i}}{\sum_i e^{-\beta E_i}}. \quad ...(v)$$

Applications : The canonical distribution is very important in statistical mechanics and can be applied to discuss many cases of physical interest.

It can be used to calculate the mean value of various characteristic parameters of a system which is in thermal contact with its environment (heat-reservoir). If a parameter y of a system has a value y_i in the state i of the system, then the mean value of y is given by

$$\bar{y} = \sum_i P_i y_i = \frac{\sum_i y_i e^{-\beta E_i}}{\sum_i e^{-\beta E_i}}.$$

where the summation is over all states i of the system.

It can thus be used to compute the mean energy or mean pressure of a gas and to investigate the magnetic properties of a substance placed in a magnetic field.

Partition Function

The partition function 'Z ' of a system is the sum of all the Boltzmann factors over all energy states of the system:

$$Z = \sum_i e^{-\beta E_i}.$$

Let us consider a dilute ideal gas in equilibrium at an absolute temperature T. Let us focus attention on a particular molecule of the gas and regard it as a small system in thermal contact with a heat reservoir (at temperature T) consisting of all the other molecules of the gas. The probability of finding the molecule in any one of its quantum states i, where its energy is E_i, is given by the canonical distribution

$$P_i = \frac{e^{-\beta E_i}}{\sum_i e^{-\beta E_i}} \qquad \text{where } \beta = \frac{1}{kT}$$

Now, if a molecule is found with probability P_i in a state i of energy E_i, then its mean energy is given by

$$\overline{E} = \Sigma P_i E_i = \frac{\sum\limits_i E_i e^{-\beta E_i}}{\sum\limits_i e^{-\beta E_i}} \qquad \text{...(i)}$$

where the summation is over all possible states *i* of the molecule.

Let us consider the numerator of eq. (i).

$$\sum_i E_i e^{-\beta E_i}. = -\sum_i \frac{\partial}{\partial \beta}\left(e^{-\beta E_i}\right) = -\frac{\partial}{\partial \beta}\left(\sum_i e^{-\beta E_i}\right)$$

Making this substitution in eq. (i), we get

$$\overline{E} = \frac{-\frac{\partial}{\partial \beta}\left(\sum\limits_i e^{-\beta E_i}\right)}{\sum\limits_i e^{-\beta E_i}}$$

Let us now define the quantity $\sum\limits_i e^{-\beta E_i} \equiv Z$. Then, we get

$$\overline{E} = \frac{-\frac{\partial Z}{\partial \beta}}{Z} = -\frac{1}{Z}\frac{\partial Z}{\partial \beta}$$

or $$\overline{E} = -\frac{\partial}{\partial \beta}\log_e Z.$$

The calculation of mean energy $\overline{E}$ thus requires only the evaluation of the sum Z . The sum Z over all states of the molecule is the 'partition function' of the molecule.

Again, let us consider an assembly of molecules in equilibrium at absolute temperature *T*. The probability of finding a molecule in a particular microstate of energy E_i is

$$P_i = \frac{e^{-\beta E_i}}{\sum\limits_i e^{-\beta E_i}}, \qquad \text{where } \beta = \frac{1}{kT}.$$

Now, the denominator $\sum\limits_i e^{-\beta E_i}$ is the partition function Z of the molecule.

$$\therefore \qquad P_i = \frac{e^{-E_i/kT}}{Z}$$

By probability concept

$$P_i = \frac{n_i}{N}$$

where n_i is the number of molecules having energy E_i and N is the total number of molecules.

From the last two expressions

$$n_i = \frac{N}{Z} e^{-E_i/kT}$$

The graph between n_i and E_i will be an exponential curve.

***Law of Equipartition of Energy* :** This law states : *If a system obeying classical statistical mechanics is in equilibrium at absolute temperature T, then every independent quadratic term in its energy expression has a mean value equal to $\frac{1}{2}kT$, where k is Boltzmann's constant.**

Let us consider a system with f degrees of freedom. Classically, this system is described by f position coordinates $q_1, q_2, \ldots, q_f$ and f corresponding momenta coordinates $p_1, p_2, \ldots, p_f$. Its energy E is then a function of these variables, that is,

$$E = E(q_1, q_2, \ldots, q_f ; p_1, p_2, \ldots, p_f).$$

Let p_i be any particular momentum. Then, the total energy can be written in the form

$$E = E_i(P_i) + E'(q_1, q_2, \ldots, q_f \; ; \; p_1, p_2, \ldots, p_f, \text{ excluding } p_i) \qquad \ldots\text{(i)}$$

where E_i is a function of p_i only while E' depends on all the variables *except* P_i.

Suppose that the system is in thermal equilibrium with its surroundings (heat reservoir) at an absolute temperature T.

The canonical distribution for the system, being a function of all the *continuous* variables, can be written as

$$P(q_1, q_2 \ldots, q_f) = \frac{e^{-\beta E(q_1, \ldots, p_f)}}{\int e^{-\beta E(q_1, \ldots, p_f)}\, dq_1, \ldots, dp_f},$$

where $\beta = 1/kT$. This is the probability of finding the system with its position and momenta coordinates in a range near $(q_1, \ldots, q_f;\, p_1, \ldots, p_f)$. By definition, the mean value of E_i is given by

$$\bar{E}_i = \frac{\int E_i\, e^{-\beta E(q_1, \ldots, p_f)}\, dq_1, \ldots, dp_f}{\int e^{-\beta E(q_1, \ldots, p_f)}\, dq_1, \ldots, dp_f},$$

where the integrals extend over all possible values of $q_1, \ldots, q_f$ and $p_1, \ldots, p_f$. *In* view of *eq.* (i), the last eq. may be written as

$$\bar{E}_i = \frac{\int E_i\, e^{-\beta(E_i + E')}\, dq_1, \ldots, dp_f}{\int e^{-\beta(E_i + E')}\, dq_1, \ldots, dp_f}$$

$$= \frac{\int E_i\, e^{-\beta E_i}\, dp_i \int e^{-\beta E'}\, dq_1, \ldots, dp_f\, (\text{except}\, dp_i)}{\int e^{-\beta E_i}\, dp_i \int e^{-\beta E'}\, dq_1, \ldots, dp_f\, (\text{except}\, dp_i)}$$

Such a separation of integrals is possible because E' is independent of p_i. The second integrals in numerator and denominator cancel mutually. Thus

$$\bar{E}_i = \frac{\int E_i\, e^{-\beta E_i}\, dp_i}{\int e^{-\beta E_i}\, dp_i}$$

$$= \frac{-\dfrac{\partial}{\partial \beta}\left[\int e^{-\beta E_i}\, dp_i\right]}{\int e^{-\beta E_i}\, dp_i}$$

$$= -\frac{\partial}{\partial \beta} \log_e \left[\int e^{-\beta E_i}\, dp_i\right]$$

The limits of the integral may be taken from $-\infty$ to $+\infty$ as the momentum p_i can assume all possible values from $-\infty$ to $+\infty$. Thus

$$\overline{E}_i = -\frac{\partial}{\partial\beta}\log_e\left[\int_{-\infty}^{+\infty} e^{-\beta E_i}\,dp_i\right] \quad \text{...(ii)}$$

In order to evaluate the integral in eq. (ii) we have to assume some form for E_i as a function of p_i . Let E_i be a quadratic function of p_i as it would be if it represents a kinetic energy (we know that K.E.= $p^2/2m$). Let us write

$$\overline{E}_i = bp_i^2,$$

where b is some constant. Then, the integral in eq. (ii) takes the form

$$\int_{-\infty}^{+\infty} e^{-\beta E_i}\,dp_i = \int_{-\infty}^{+\infty} e^{-\beta bp_i^2}\,dp_i$$

$\sqrt{\beta}\,p_i$ so that $dy = \sqrt{\beta}\,dp_i$, we obtain

$$\int_{-\infty}^{+\infty} e^{-\beta E_i}\,dp_i = \frac{1}{\sqrt{\beta}}\int_{-\infty}^{+\infty} e^{-by^2}\,dy$$

Making this substitution in eq. (ii), we get

$$\overline{E}_i = -\frac{\partial}{\partial\beta}\log_e\left[\frac{1}{\sqrt{\beta}}\int_{-\infty}^{+\infty} e^{-by^2}\,dy\right]$$

$$= -\frac{\partial}{\partial\beta}\left[-\frac{1}{2}\log_e\beta + \log_e\int_{-\infty}^{+\infty} e^{-by^2}\,dy\right]$$

The derivative of $\log_e\int_{-\infty}^{+\infty} e^{-by^2}\,dy$ with respect to β is zero because it, being definite integral, does not contain β explicitly. Hence we get

$$\therefore \quad \overline{E}_i = -\frac{\partial}{\partial\beta}\left[-\frac{1}{2}\log_e\beta\right] = \frac{1}{2\beta}$$

or

$$\boxed{\overline{E}_i = \frac{1}{2}kT,} \qquad \left[\because \beta = \frac{1}{kT}\right]$$

which is the law of equipartition of energy.

Limitation **:** The law of equipartition of energy is applicable only so long the system can be described classically, usually at high temperatures.

Application : Let us apply the law to evaluate the specific heat of a monatomic ideal gas. The energy of a molecule in such a gas is simply its kinetic energy, *i.e.,*

$$E_i = \frac{(px^2 + py^2 + pz^2)}{2m} = \frac{px^2}{2m} + \frac{py^2}{2m} + \frac{pz^2}{2m}$$

By equipartition law, the mean value of each of the three terms in this expression is $\frac{1}{2}kT$, where T is absolute temperature of the gas.

$$\therefore \quad \bar{E}_i = (3/2)\,kT\,.$$

Since one mole of gas contains N_A (Avogadro's number) molecules, the mean energy of the gas (per mole) is

$$\bar{E} = N_A\left(\frac{3}{2}kT\right) = \frac{3}{2}RT\,,$$

where R (- N_A k) is the gas constant. Therefore, the molar specific heat at constant volume is given by

$$C_V = \left(\frac{\partial \bar{E}}{\partial T}\right)_V = \frac{2}{3}R.$$

Principle of Equipartition of Energy

It states that *if a classical dynamical system is in thermal equilibrium at the absolute temperature T, then every independent quadratic term in its energy has a mean value equal to* $\frac{1}{2}kT$.

Mean Energy of a Harmonic Oscillator : Let us consider a particle of mass m performing simple harmonic oscillations in one dimension. Its energy is then given by

$$E = \frac{p_x^2}{2m} + \frac{1}{2}\alpha x^2 \qquad \text{...(i)}$$

The first term is the kinetic energy of the particle in terms of its momentum p_x, and the second term is its potential energy, α is the force constant. This expression of energy has two quadratic terms.

Suppose that the oscillator is in equilibrium with a heat reservoir at a temperature T high enough so that we can describe the oscillator in terms of classical mechanics.

The mean energy of this oscillator is given by

$$\overline{E} = \frac{\int E e^{-\beta E} dx\, dp_x}{\int e^{-\beta E} dx\, dp_x} \qquad \text{where } \beta = 1/kT$$

and the integrals extend over all possible values of x and p_x ($-\infty$ to $+\infty$).

Using eq. (i), the above expression can be written as

$$\overline{E} = \frac{\int \frac{p_x^2}{2m} e^{-\beta p_x^2/2m} dp_x}{\int e^{-\beta p_x^2/2m} dp_x} + \frac{\int \frac{1}{2}\alpha x^2 e^{-\beta\alpha x^2/2} dx}{\int e^{-\beta\alpha x^2/2} dx}$$

$$= \frac{-\frac{\partial}{\partial\beta}\left[\int e^{-\beta p_x^2/2m} dp_x\right]}{\int e^{-\beta p_x^2/2m} dp_x} + \frac{-\frac{\partial}{\partial\beta}\int e^{-\beta\alpha x^2/2} dx}{\int e^{-\beta\alpha x^2/2} dx}$$

$$= -\frac{\partial}{\partial\beta}\log_e\left[\int_{-\infty}^{+\infty} e^{-\beta p_x^2/2m} dp_x\right] - \frac{\partial}{\partial\beta}\log_e\left[\int_{-\infty}^{+\infty} e^{-\beta\alpha x^2/2} dx\right]$$

Putting $y = \sqrt{\beta} p_x$ and $z = \sqrt{\beta} x$ in the first and second integrals respectively, we can show that

$$\overline{E} = -\frac{\partial}{\partial\beta}\left[-\frac{1}{2}\log_e \beta + \log_e \int_{-\infty}^{+\infty} e^{-y^2/2m} dy\right]$$

$$-\frac{\partial}{\partial\beta}\left[-\frac{1}{2}\log_e \beta + \log_e \int_{-\infty}^{+\infty} e^{-\alpha z^2/2} dz\right]$$

The derivatives of $\log_e \int_{-\infty}^{+\infty} e^{-y^2/2m} dy$ and $\log_e \int_{-\infty}^{+\infty} e^{-\alpha z^2/2} dz$ are each zero because they, being definite integrals, do not contain β explicitly. Hence we get

$$\bar{E} = -\frac{\partial}{\partial\beta}\left[-\frac{1}{2}\log_e \beta\right] - \frac{\partial}{\partial\beta}\left[-\frac{1}{2}\log_e \beta\right]$$

$$= \frac{1}{2\beta} + \frac{1}{2\beta}$$

$$= \frac{1}{2}kT + \frac{1}{2}kT = kT.$$

This result is obvious from the principle of equipartition of energy. The energy equation (i) has two independent quadratic terms and each of them must have a mean value equal to $\frac{1}{2}kT$.

PROBLEMS

1. In a system in thermal equilibrium at absolute temperature T, two states with energy difference 4.83 x 10^{-21} joule occur with relative probability e^2. Deduce the temperature (k = 1.38 x 10^{-23} joule/K).

Solution : Using canonical distribution, the probability of finding the system in a particular state *i* of energy E_i is given by

$$P_i = Ce^{-\beta E_i},$$

where C is a proportionality constant and $\beta = 1/kT$. Now, if the system has two states of energies E_1 and E_2, then

$$P_1 = Ce^{-\beta E_1} \text{ and } P_2 = Ce^{-\beta E_2}$$

$$\therefore \quad \frac{P_1}{P_2} = \frac{e^{-\beta E_1}}{e^{-\beta E_2}} = e^{\beta(E_2 - E_1)}$$

or $$\log_e\left(\frac{P_1}{P_2}\right) = \beta(E_2 - E_1) = (E_2 - E_1)/kT$$

or $$T = \frac{(E_2 - E_1)}{k \log_e (P_1 / P_2)}$$

Here $E_2 - E_1 = 4.83 \times 10^{-21}$ joule

and $\log_e (P_1/ P_2) = \log_e(e^2) = 2$. Therefore,

$$T = \frac{4.83 \times 10^{-21} \text{ joule}}{(1.38 \times 10^{-23} \text{ joule/K}) \times 2} = 175\,K.$$

2. An excited state of an atom is 1-38 eV above the ground state. Calculate the number of atoms in this excited state relative to the ground state at 16000 K.(k = 1.38x10^{-23} joule/K).

Solution : Let E_1 and E_2 be the energies of the ground state and the excited state. By canonical distribution, the probabilities of finding an atom in these states are

$$P_1 = Ce^{-\beta E_1}$$

and $$P_2 = Ce^{-\beta E_2},$$

where C is a constant and $\beta = \frac{1}{kT}$.

$$\therefore \quad \frac{P_2}{P_1} = \frac{e^{-\beta E_2}}{e^{-\beta E_1}} = e^{-\beta(E_2 - E_1)}$$

$$= e^{-(E_2 - E_1)/kT}$$

$$= \exp\left[-\frac{1.38 \times (1.6 \times 10^{-19})\text{ J}}{(1.38 \times 10^{-23}\text{ J/K}) \times 16000\text{ K})}\right]$$

$$= e^{-1}$$

$$= 0.368.$$

This is the fraction of atoms in the excited state relative to the ground state.

3. Estimate the temperature of a gas containing hydrogen atoms at which the Balmer series lines will be observed in the absorption spectrum. The energies of the ground state and the

first excited state of hydrogen atom are E_1 = - 13.6 eV and E_2 = -3.39 eV respectively.

($k = 1.38 \times 10^{-23}$ joule/K = 8.62×10^{-5} eV/K)

Solution : Balmer lines in absorption can be observed provided a significant fraction of the hydrogen atoms are initially in the first excited state.

According to Boltzmann probability distribution, the ratio of the number n_2 of atoms in the first excited state to the number n_1 of atoms in the ground state, in a large sample of gas in equilibrium at absolute temperature T, is

$$\frac{n_2}{n_1} = \frac{e^{-E_2/kT}}{e^{-E_1/kT}}$$

$$= e^{-(E_2-E_1)/kT}$$

$$= e^{-(-3.39+13.6)eV/(8.62\times 10^{-5}eV/k)T}$$

$$= e^{-1.18\times 10^{5}k/T}$$

Clearly, for a significant fraction n_2/n_1, the temperature T should be of the order of 10^5 K.

4. Each atom of a gas has only two states, one of energy zero and the other of energy 4.6 eV. What fraction of the atoms will be in the lower state at temperature (i) 300 K, (ii) 30,000 K. Take $k = 1.38 \times 10^{-23}$ joule/K.

Solution : The probability of finding an atom in a state i of energy E_i is given by

$$P_i = C e^{-E_i/kT}$$

where C is a constant of proportionality.

There are only two energy states, $E_1 = 0$ and $E_2 = 4.6$ eV. The probabilities of finding the atom in these states are

$$P_1 = C e^{-0} = C \qquad ...(i)$$

and $$P_2 = C e^{-4.6\,eV/kT(\text{in }eV)} \qquad ...(ii)$$

Since the exponential term is less than 1, the state of lower energy (E_1= 0) is more probable than the state of higher energy (E_2 = 4.6 eV), as expected.

The constant C can be determined by the normalization condition that the probability of finding the atom in either one of these states must be unity. That is

$$P_1 + P_2 = C + Ce^{-4.6/kT} = 1$$

or
$$C = \frac{1}{1+e^{-4.6/kT}} \qquad \text{...(iii)}$$

By definition, the probability P_i denotes the fraction of atoms in a given state E_i. Thus :

(i) At T = 300 K, the fraction of atoms in the lower state is, using (i) and (iii)

$$P_1 = C = \frac{1}{1+\exp\left[-\frac{4.6 \times (1.6 \times 10^{-19})}{(1.38 \times 10^{-23}) \times 300}\right]}$$

[∵ 1 eV = 1.6 x 10^{-19} joule]

$$= \frac{1}{1+e^{-178}}$$

$$\simeq 1 \text{(almost all atoms).} \qquad [\because e^{-178} \simeq 5 \times 10^{-78}]$$

(ii) At T = 30,000 K, the fraction of atoms in the lower state is

$$P_1' = C = \frac{1}{1+\exp\left[-\frac{4.6 \times (1.6 \times 10^{-19})}{(1.38 \times 10^{-23}) \times 300}\right]}$$

$$= \frac{1}{1+e^{-1.78}}$$

$$= 0.85 \qquad [\because e^{-1.78} \simeq 0.17]$$

5. (a) A system can take only three different energy states $E_1=0$, $E_2 = 1.38 \times 10^{-21}$ joule, $E_3 = 2.76 \times 10^{-21}$ joule. These states occur in 2, 5, 4 different ways respectively. Deduce the probability that at temperature 100 K the system may be (i) in one of the microstates of energy E_3 and (ii) in ground state E_1. ($k = 1.38 \times 10^{-23}$ joule/K).

(b) Calculate the average energy of the system at 100 K.

Solution : (a) The probability P_i of finding a system in thermal equilibrium in any one particular state i of energy E_i is given by

$$P_i = \frac{e^{-\beta E_1}}{\sum_i e^{-\beta E_1}} \qquad \beta = \frac{1}{kT}$$

The given three states E_1, E_2, E_3 occur in 2, 5, 4 different ways. The probability distributions of these states are

$$g_1 e^{-E_1/kT} = 2e^{-0/kT} = 2e^{-0} = 2,$$

$$g_2 e^{-E_2/kT} = 5\exp\left[-\frac{1.38 \times 10^{-21}}{1.38 \times 10^{-23} \times 100}\right] = 5e^{-1} = 5 \times 0.368$$

$$= 1.84$$

and

$$g_3 e^{-E_3/kT} = 4\exp\left[-\frac{2.76 \times 10^{-21}}{1.38 \times 10^{-23} \times 100}\right] = 4e^{-2} = 4 \times 0.315$$

$$= 0.54$$

(i) The probability of the system for being in one of the microstates of E_3 is

$$P_3 = \frac{g_3 e^{-E_3/kT}}{\sum_{i=1}^{3} g_i e^{-E_i/kT}}$$

$$= \frac{0.54}{2+1.84+0.54} = \frac{0.54}{4.38} = 0.123.$$

(ii) The probability for the ground state is

$$P_1 = \frac{g_1 e^{-E_1/kT}}{\sum_{i=1}^{3} g_i e^{-E_i/kT}}$$

$$= \frac{2}{2+1.84+0.54} = \frac{2}{4.38} = 0.457.$$

(b) The average energy of the system at 100 K is given by

$$\overline{E} = \frac{\sum_i E_i g_i e^{-E_i/kT}}{\sum_i g_i e^{-E_i/kT}}$$

$$= \frac{(0 \times 2) + (1.38 \times 10^{-21} \times 1.84) + (2.76 \times 10^{-21} \times 0.54)}{2+1.84+0.54}$$

$$= \frac{0 + (2.45 \times 10^{-21}) + (1.49 \times 10^{-21})}{2+1.84+0.54}$$

$$= \frac{4.03 \times 10^{-21}}{4.38} = 0.92 \times 10^{-21} \text{ joule.}$$

6. *A system can exist in three energy states $E_1 = 0$, $E_2 = 1.4 \times 10^{-14}$ erg and $E_3 = 2.8 \times 10^{-14}$ erg. These states occur in 2, 5 and 4 different ways respectively. Find the probabilities of the system of being in these states. ($k = 1.4 \times 10^{-16}$ erg/K)*

0.457, 0.420, 0.123.

5

Spontaneous Flow

Heat Conduction

Whenever two parts of a body are at different temperatures, a spontaneous flow of heat takes place from the region of higher temperature to that of lower temperature. This process is known as 'conduction' of heat. It can be explained on a microscopic basis. Microscopically, heat is mechanical energy of individual atoms and molecules. When the molecules in one region have, on the average, more kinetic energy than those in a neighbouring region, they transfer energy to their neighbours by colliding with them. This transfer of energy takes place from a region of higher temperature, and hence greater molecular motion, to one of lower temperature.

In metals, the conduction of heat takes place by a different agency also. Metals have some "free" electrons which are able to move throughout the material, rather than being bound tightly to individual atoms. These electrons carry energy readily from the hotter to the colder region. In fact, this is the chief mechanism of heat conduction in metals. It is for this reason that materials which are good conductors of electricity are generally also good conductors of heat.

Heat Conduction and Wave Propagation : In heat conduction, energy is transferred from one part of a medium to the other by virtue of a temperature-difference between them, and the direction of energy flow is always from points of higher to points of lower temperature. In wave propagation, the energy is transferred from one part of a material medium to the other by virtue of the elasticity and inertia of the medium, and flows *in all directions* .

Steady State : Let us take a bar of uniform cross-section and heat it at one end. Each cross-section of the bar receives heat by conduction from the adjacent cross-section towards the heated end. This heat is spent in three ways ; a part is absorbed by the cross-section to increase its temperature, another part escapes out from the sides of the cross-section, and the third is conducted to the next cross-section. The same happens at the next cross-section, and so on. Thus the temperature of each part of the bar rises, and the bar is said to be in a "variable state. After some time, however, *a state is reached when temperature at each point of the bar becomes stationary*. This is called the 'steady state'. In this state *no* heat is absorbed by the bar. The heat that reaches any section is transferred to the next, except that some heat may escape out from the sides. (The steady state is not the same as thermal equilibrium, in which all parts of the bar must have the *same* temperature).

The steady state is tried to be achieved in an experiment in which the heat supplied to a body in a given time is to be measured, such as Searle's experiment for determining the thermal conductivity of a rod. In this experiment the heat supplied at one end of the rod is determined by measuring the rise in temperature of a known amount of water flowing around the other end. The measurement is done after the steady state is reached because it is only then that all the heat supplied at one end reaches the other end where it is taken in by water. Before the steady state, a part of the heat supplied is used up in increasing the temperature of the rod and hence the heat taken in by water is not equal to the heat supplied to the rod.

Steady Heat-flow and Steady Fluid-flow : The steady heat-flow is analogous to the steady flow of an incompressible fluid. In the steady heat-flow, the temperature at each point of the medium is 'stationary' and the rate of heat-flow is the same. In the steady flow of an incompressible fluid, the pressure at each point of the fluid is stationary and the velocity of each passing fluid particle is always the same.

Coefficient of Thermal Conductivity : Let us consider a parallel-faced slab whose cross-sectional area is A and thickness *d*. Let its faces be maintained at steady temperatures θ_1 and θ_2. Heat will flow through the slab from the face at higher temperature to that at the lower temperature. If it is assumed that no heat escapes out from the sides of the slab, the lines of flow will be at all points normal to the faces of the slab. In these circumstances, under steady state, the quantity of heat *Q* entering one face in a given time is same as the heat leaving the other face in the same time. Experiment shows that *Q* is proportional to (i) the area *A* , (ii) temperature difference $\theta_1 - \theta_2$ between the faces, (iii) time of flow *t*, and (iv) inversely proportional to the thickness *d*, that is

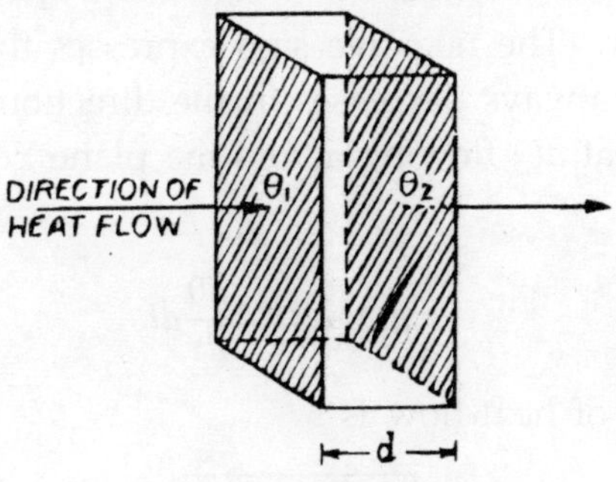

$$Q \propto A \frac{\theta_1 - \theta_2}{d}$$

or $$Q = KA\frac{\theta_1 - \theta_2}{d}t, \qquad \text{...(i)}$$

where K is a constant depending on the *material* of the slab. It is called the "coefficient of thermal conductivity" of the material of the slab. The quantity $\frac{\theta_1 - \theta_2}{d}$ is the 'temperature-gradient' across the slab.

If $A = 1, \frac{\theta_1 - \theta_2}{d} = 1$ and $t = 1$, then $K = Q$. Therefore, *the coefficient of thermal conductivity of a material is defined as the amount of heat that flows in unit time through unit area of the material perpendicular to the flow under unit temperature-gradient, when the steady state has been reached*. Its units are cal-cm^{-1} – sec^{-1} –°C^{-1}, and dimensions [MLT^{-1} Θ^{-1}].

A material for which K is large is a good heat conductor, while if K is small, the material is a poor conductor or a good insulator.

In case when the heat escapes out from the sides of the slab, the temperature-gradient is not uniform across the slab. In such a case, if θ and $\theta + \theta 0$ be the temperatures at planes distant x and $x + \partial x$ along the direction of heat-flow, then the temperature-gradient will be $\frac{\theta - (\theta + \delta\theta)}{\delta x} = -\frac{\delta\theta}{\delta x}$. In the limit $\delta x \to 0$; this becomes $-\, d\theta/dx$ which is the temperature-gradient *at the plane x*. (The negative sign expresses the fact that the temperature always decreases in the direction of heat-flow.) Then, the heat dQ flowing across the plane x in time dt will be given by

$$dQ = -KA\frac{d\theta}{dx}dt.$$

The rate of heat-flow is

$$\boxed{\frac{dQ}{dt} = -KA\frac{d\theta}{dx}.}$$

Heat Sensation from Metal and Wood : Our sensation of the hotness or coldness of the substances that we touch depend upon their conductivity. In hot summer, a piece of iron appears much hotter than a piece of wood although they are both at the same temperature. This is because iron is a good heat-conductor but wood is a poor conductor. In summer , the temperature of our body is lower than the temperature of the iron or wood. Therefore, when we touch iron, heat readily flows from *all* parts of the iron to our hand and we feel hot. But when we touch wood, heat is conducted to our hand only from the part of the wood in contact with our hand and we do not feel so hot. Similarly, on a cold winter day, when the temperature of the body is higher than that of iron or wood, the iron feels colder then the wood. In this case the flow of heat takes place from the hand to the iron wood. When we touch iron we *continuously* give heat to it who spreads it rapidly to all its parts and hence we feel cold. The wood cannot do so and as such little heat flows from the hand to the wood.

If the iron or wood are *at the same temperature as the temperature of our body,* no heat-transference will take place when they are touched, and they will feel both equally hot or equally cold.

Heat and Temperature: Heat conduction shows that heat and temperature are quite different conceptions. Different bars, having same temperature-difference between their ends, may transfer *different* quantities of heat in the same time.

Thermoelectric Conductivity (or Diffusivity) : It is defined as the ratio of the coefficient of thermal conductivity to the thermal capacity per unit volume of the material. It p be the density of the material and *c* its specific heat, then thermal capacity per unit volume is *pc*. Hence, if *K* be the thermal conductivity, the thermometric conductivity will be given by

$$k = \frac{K}{\rho c}$$

When one end of a bar is heated, the thermoelectric conductivity determines the rate at which the temperature of any part of the bar rises, before the steady state is reached.

The rate at which the temperature of a body rises (or falls) when heated (or cooled) depends upon the thermal capacity of the body, and is smaller when the thermal capacity is higher. But once a body attains a certain temperature then the sensation of hotness or coldness which it produces when touched depends only on the thermal conductivity of the body.

Cooking Utensils : A cooking utensil should be of a material of large thermal conductivity, low specific heat and low coefficient of expansion.

Thermal Resistance : Just as charge flows in an electrical circuit due to a potential difference between two points of the circuit, in the same way heat flows in a conductor due to a temperature difference between two points of the conductor. Hence, like electrical resistance, there is also a thermal resistance in a material.

Let l be the length and A the area of cross-section of a rod. Let θ_1 and θ_2 be the temperatures , of the hot and the cold ends of the rod in the steady state. Then, the rate of flow of heat in the rod is given by

$$H = \frac{Q}{t} = KA\frac{\theta_1 - \theta_2}{l}$$

or
$$\frac{\theta_1 - \theta_2}{H} = \frac{l}{KA}$$

$(\theta_1 - \theta_2)$ is the temperature-difference between the ends of the conductor and H is the rate of flow of heat. $(\theta_1 - \theta_2)/H$ is called the 'thermal resistance' R of the conductor. Thus

$$R = \frac{\theta_1 - \theta_2}{H} = \frac{l}{KA}$$

Thus, greater the coefficient of thermal conductivity of a material, smaller is the thermal resistance of a rod of that material. The unit of thermal resistance is 'second-°C/calorie'.

Let us consider two slabs of thicknesses d_1, d_2 and conductivities K_1, K_2. Let θ_1 and θ_2 be the temperatures of the end faces and θ the temperature of the interface. The rates of flow of heat through the thicknesses d_1 and d_2 are given by

$$\frac{dQ_1}{dt} = K_1 A \frac{\theta_1 - \theta}{d_1} \qquad \text{...(i)}$$

and
$$\frac{dQ_2}{dt} = K_2 A \frac{\theta - \theta_2}{d_2} \qquad \text{...(ii)}$$

In the *steady* state, the rate of flow of heat throughout the compound slab will be the same *i.e.*, $\frac{dQ_1}{dt} = \frac{dQ_2}{dt}$, so that

$$K_1 A \frac{\theta_1 - \theta}{d_1} = K_2 A \frac{\theta - \theta_2}{d_2}$$

or
$$\frac{K_1\theta_1}{d_1} + \frac{K_2\theta_2}{d_2} = \left\{\frac{K_1}{d_1} + \frac{K_2}{d_2}\right\}\theta$$

$\therefore$
$$\theta = \frac{\dfrac{K_1\theta_1}{d_1} + \dfrac{K_2\theta_2}{d_2}}{\dfrac{K_1}{d_1} + \dfrac{K_2}{d_2}}$$

This is the expression for the temperature of the interface.

Let dQ/dt be the rate of heat-flow through the compound slab. Then, substituting the above value of θ in eq. (i), we get

$$\frac{dQ}{dt} = \frac{K_1 A}{d_1}\left[\theta_1 - \frac{\dfrac{K_1\theta_1}{d_1} + \dfrac{K_2\theta_2}{d_2}}{\dfrac{K_1}{d_1} + \dfrac{K_2}{d_2}}\right]$$

$$= \frac{K_1 A}{d_1}\left[\frac{\frac{K_1\theta_1}{d_1}+\frac{K_2\theta_1}{d_1}-\frac{K_1\theta_1}{d_1}-\frac{K_2\theta_2}{d_2}}{\frac{K_1}{d_1}+\frac{K_2}{d_2}}\right]$$

$$= \frac{K_1 A}{d_1}\left[\frac{\frac{K_2}{d_d}(\theta_1-\theta_2)}{\frac{K_1}{d_1}+\frac{K_2}{d_2}}\right]$$

$$= \frac{K_1 K_2}{d_1 d_2}\left[\frac{A(\theta_1-\theta_2)}{\frac{K_1}{d_1}+\frac{K_2}{d_2}}\right]$$

$$= \frac{A(\theta_1-\theta_2)}{\frac{d_1}{K_1}+\frac{d_2}{K_2}}$$

For n slabs in series, the expression would be

$$\frac{dQ}{dt} = \frac{A(\theta_1-\theta_n)}{\Sigma(d/K)}$$

Let A be the cross-sectional area and d the thickness of each plate. Let θ_1, θ_2 and θ be the steady temperatures at the first face, second face and at the interface respectively. When the heat flows through the composite plate, its rate will be the same in each plate. Thus

$$\frac{\Delta Q}{\Delta t} = K_1 A\frac{(\theta_1-\theta)}{d} = K_2 A\frac{(\theta-\theta_2)}{d} \quad \text{...(i)}$$

or $$K_1\theta_1 - K_1\theta = K_2\theta - K_2\theta_2$$

or $$K_1\theta_1 + K_2\theta_2 = K_1\theta + K_2\theta$$

$\therefore$ $$\theta = \frac{K_1\theta_1+K_2\theta_2}{K_1+K_2} \quad \text{...(ii)}$$

If K be the equivalent thermal conductivity, then we have

$$\frac{\Delta Q}{\Delta t} = KA\frac{\theta_1 - \theta_2}{2d} \qquad \ldots\text{(ii)}$$

Equating it to (i), we get

$$KA\frac{\theta_1 - \theta_2}{2d} = K_1 A\frac{\theta_1 - \theta}{d}$$

or

$$K = 2K_1\frac{\theta_1 - \theta}{\theta_1 - \theta_2}$$

Substituting the value of θ from eq. (ii), we get

$$K = \frac{2K_1}{\theta_1 - \theta_2}\left[\theta_1 - \frac{K_1\theta_1 + K_2\theta_2}{K_1 + K_2}\right]$$

$$= \frac{2K_1}{\theta_1 - \theta_2}\left[\frac{K_1\theta_1 + K_2\theta_1 - K_1\theta_1 - K_2\theta_2}{K_1 + K_2}\right]$$

$$= \frac{2K_1}{\theta_1 - \theta_2}\frac{K_2(\theta_1 - \theta_2)}{K_1 + K_2}$$

$$= \frac{2K_1K_2}{K_1 + K_2}$$

This is the required expression.

The thermal conductivity of a perfect heat conductor is infinite.

Let A be the cross-sectional area of each slab. Let θ_1, θ_2, θ_3, θ_4 be the steady temperatures at the first face, 1-2 interface, 2-3 interface and the third face respectively. The rate of heat flow would be the same throughout. Thus

$$K_1 A\frac{\theta_1 - \theta_2}{d_1} = K_2 A\frac{\theta_2 - \theta_3}{d_2}$$

$$= K_3 A\frac{\theta_3 - \theta_4}{d_3} = \text{constant.}$$

This gives

$$\theta_1 - \theta_2 = \text{const.}\frac{d_1}{K_1}$$

$$\theta_2 - \theta_3 = \text{const.}\frac{d_2}{K_2}$$

$$\theta_3 - \theta_4 = \text{const.}\frac{d_3}{K_3}$$

Adding:
$$\theta_1 - \theta_4 = \text{const.}\left[\frac{d_1}{K_1} + \frac{d_2}{K_2} + \frac{d_3}{K_3}\right] \quad ....(i)$$

If K be the equivalent conductivity for the combination, we may write

$$KA\frac{\theta_1 - \theta_4}{d_1 + d_2 + d_3} = \text{Const.}$$

or
$$\theta_1 - \theta_4 = \text{const.}\frac{d_1 + d_2 + d_3}{K} \quad ...(ii)$$

Comparing eq. (i) and (ii), we get

$$\frac{d_1 + d_2 + d_3}{K} = \frac{d_1}{K_1} + \frac{d_2}{K_2} + \frac{d_3}{K_3}$$

Rectilinear Flow of Heat along a Metal Bar : Let us consider a long metal bar, one end of which is maintained at a steady *excess* temperature of θ_0 above the surroundings. In the beginning, the temperatures at various points along the bar gradually rise but finally a steady state is reached at which these temperatures become stationary. A steady flow of heat now takes place along the bar. If the bar is exposed to the surroundings, a portion of the heat passing across any cross-section of the bar escapes out from the sides. Hence smaller and smaller amounts of heat pass across successive sections.

Let us consider two cross-sections of the bar, L and M, distant x and $x + \delta x$ from the hot end. Now, *before the steady state is reached,* the temperature at any section of the bar is rising. Let θ be the excess of temperature at the section

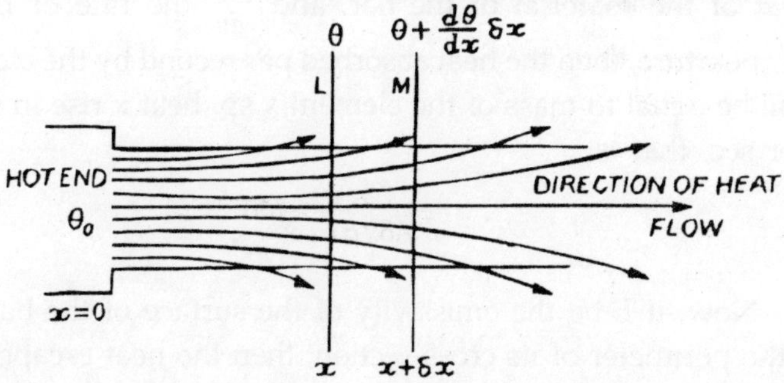

(Fig. 3)

L at any instant *t*. Then, if $-\frac{d\theta}{dx}$ be the temperature-gradient at *L* , the excess temperature at *M* will be $\theta+\frac{d\theta}{dx}\delta x$. If *A* is the area of cross-section of the bar (assumed uniform) and *K* the thermal conductivity of its material, the rate of heat-flow across the section *L* will be

$$-KA\frac{d\theta}{dx}$$

Similarly, the rate of heat-flow across the section *M* will be

$$-KA\frac{d}{dx}\left\{\theta+\frac{d\theta}{dx}\delta x\right\}$$

Therefore, the excess of heat flowing per second across *L* over that flowing across *M* is

$$-KA\frac{d\theta}{dx}-\left[-KA\frac{d}{dx}\left\{\theta+\frac{d\theta}{dx}\delta x\right\}\right]$$

$$=KA\frac{d^2\theta}{dx^2}\delta x. \qquad \text{...(i)}$$

This amount of heat is partly absorbed by the element of the bar between *L* and *M*, and partly lost from the sides of the element to the surroundings. If ρ be the density, c the specific heat of the material of the bar, and $\frac{d\theta}{dt}$ the rate of rise in temperature, then the heat absorbed per second by the element will be equal to mass of the element x sp. heat x rise in temp, per sec, that is.

$$A\delta x \rho \times c \times \frac{d\theta}{dt}. \qquad \text{...(ii)}$$

Now, if E be the emissivity of the surface of the bar and p the perimeter of its cross-section, then the heat escaped out per second from the sides of the element between L and M will be

$$E \times p \delta x \times \theta \qquad \text{...(iii)}$$

Eq. (i), (ii) and (iii) give

$$KA\frac{d^2\theta}{dx^2}\delta x = A\delta x p \times c \times \frac{d\theta}{dt} + E \times p\delta x \times \theta$$

or

$$\frac{K}{\rho c}\left\{\frac{d^2\theta}{dx^2}\right\} = \frac{d\theta}{dt} + \frac{Ep}{A\rho c}\theta$$

or

$$\frac{d\theta}{dt} = \frac{K}{\rho c}\left\{\frac{d^2\theta}{dx^2}\right\} - \frac{p}{A\rho c}\theta \qquad \text{...(iv)}$$

This is the *Fourier 's differential equation* for one-dimensional flow of heat.

Temperature Distribution along the Bar when Exposed: After the steady state is reached, the temperature at every point of the bar becomes *stationary i.e.* $\frac{d\theta}{dt} = 0$. Eq. (iv) then becomes

$$\frac{K}{\rho c}\left\{\frac{d^2\theta}{dx^2}\right\} - \frac{Ep}{A\rho c}\theta = 0$$

or $$\frac{d^2\theta}{dx^2} = \theta^2 ae^{\alpha x} \qquad \text{...(v)}$$

where $\mu^2 = Ep/KA$

Let us now solve this differential equation.. Let a solution be

$$\theta = ae^{\alpha x} \qquad \text{...(vi)}$$

where a and α are constants. Differentiating it twice, we obtain

$$\frac{d^2\theta}{dx^2} = \alpha^2 ae^{\alpha x}$$

Putting these values of $\frac{d^2\theta}{dx^2}$ and θ in eq. (v), we get

$$\alpha^2 ae^{\alpha e} = \mu^2 ae^{\alpha x}$$

or $$\alpha^2 = \mu^2$$

or $$\alpha = \pm\mu^2$$

Thus, the solution of eq. (v) will be obtained by putting $\alpha = \pm\ \mu$ in eq. (vi). That is, the solution is

$$\theta = ae^{\pm \mu x}$$

or the most general solution is

$$\theta = a_1 e^{+\mu x} + a_2 e^{-\mu x} \qquad \text{...(vii)}$$

where a_1 and a_2 are arbitrary constants, the values of which can be determined by the experimental conditions.

If the bar is considered to be infinitely long, its cold end will be nearly at the temperature of the surroundings, *i.e.*

$$\text{as } x \to \infty; \theta \to 0$$

Applying this condition in eq. (vii), we have

$$0 = a_1 e^{\mu\infty} + a_2 e^{-\mu\infty}$$

or $$0 = a_1 e^{\mu\infty} \qquad \left[\because e^{-\mu\infty} = 0\right]$$

$\therefore$ $$a_1 = 0.$$

Again, θ_0 is the excess-temperature above the surroundings at the hot end $(x = 0)$ of the bar, *i.e.*

at $$x = 0,\ \theta = \theta_0$$

Applying this condition and also putting $a_1 = 0$ in eq. (vii), we get

$$\theta_0 = 0 + a_2 e^0$$

$$\therefore \quad a_2 = \theta_0$$

Putting the values of a_1 and a_2 in eq. (vii), we get

$$\boxed{\theta = \theta_0 e^{-\mu x}}$$

This equation shows that, for an exposed bar, the temperature falls off exponentially from the heated.

Temperature-Distribution along the Bar when Covered : If the bar is covered with some non-conducting material to prevent losses of heat from its surface, then its emissivity E is zero. Now, in the steady state, the rate of flow of heat is same all along the bar because the heat reaching any section completely goes to the next section.

Now, as $E = 0, \mu^2 \left(= \frac{Ep}{KA}\right) = 0.$ Hence eq. (v) becomes

$$\frac{d^2\theta}{dx^2} = 0$$

Integrating:

$$\frac{d\theta}{dx} = a_1 \text{ (constant)}$$

Integrating again:

$$\theta = a_1 x + a_2, \qquad \text{...(viii)}$$

where a_1 and a_2 are constants.

Let l be the length of the bar. Suppose its cold end is very nearly at the same temperature as the surroundings. Then the experimental conditions are

at $$x = l,\ \theta = 0,$$

at $$x = 0,\ \theta = \theta_0$$

Applying these conditions in eq. (viii), we get

$$0 = a_1 l + a_2$$

and $$\theta_0 = a_2$$

$\therefore$ $$a_1 = -\frac{\theta_0}{l}$$

and $$a_2 = \theta_0$$

Putting these values in eq. (viii), we get

$$\boxed{\theta = -\frac{\theta_0}{l}x + \theta_0}$$

This equation shows that, for a covered bar, the temperature falls off linearly from the heated end.

Ingen-Hausz's Experiment

An approximate method of comparing the thermal conductivities of materials in the form of long thin rods was devised by Ingen-Hausz. The rods are taken identical in size and shape and similarly polished. They are then thinly coated with wax and placed with one end inserted in an oil-bath. The oil in the bath is maintained at its boiling point. When the steady state is reached, the wax is found to be melted off upto different lengths along the different bars.

Let l_1, l_2, l_3, ... be the lengths upto which the wax is melted along the bars 1, 2, 3,... Then, if K_1, K_2, K_3.... be their thermal conductivities, we can prove that

$$K_1 : K_2 : K_3 .... = l_1^2 : l_2^2 : l_3^2 ....$$

Proof : After the steady state is reached, the heat flows steadily along the length of each bar from the hot to the cold end. Let us concentrate our attention on any one bar. Let us consider two cross-sections of the bar distant x and $x + \delta x$ from the hot end. Let θ be the excess-temperature over the surroundings at the section at x. Then if $\frac{d\theta}{dx}$ be the temperature-gradient at x, the excess-temperature at the section at $x + \delta x$ will be $\theta + \frac{d\theta}{dx}\delta x$. If A is the area of cross-section of the bar and K the thermal conductivity of its material, then the rate of heat-flow across the section at x will be

$$-KA\frac{d\theta}{dx}.$$

Similarly, the rate of heat-flow across the section at $x + \delta x$ will be

$$-KA\frac{d}{dx}\left\{\theta + \frac{d\theta}{dx}\delta x\right\}$$

Therefore, the excess of heat flowing per second across the section at x over that flowing across the section at $x + \delta x$ is

$$-KA\frac{d\theta}{dx} - \left[-KA\frac{d}{dx}\left\{\theta + \frac{d\theta}{dx}\delta x\right\}\right]$$

$$= KA\frac{d^2\theta}{dx}\delta x$$

This amount of heat is escaping out per second from the sides of the bar between x and $x + \delta x$. Now, if E be the emissivity of the surface of the bar, and p the perimeter of its cross-section, then the heat escaped out per second from the sides of the bar between x and $x + \delta x$ will be

$$E \times p\delta x \times \theta$$

Hence, equating the last two expressions, we have

$$KA\frac{d^2\theta}{dx^2}\delta x = E\times p\,\delta x\times\theta$$

or
$$\frac{d^2\theta}{dx^2} = \frac{Ep}{KA}\theta = \mu^2\theta$$

where $\mu^2 = \frac{Ep}{KA}$. The most general solution of this differential equation will be

$$\theta = a_1e^{+\mu x} + a_2e^{-\mu x} \qquad ...(i)$$

where a_1 and a_2 are arbitrary constants, the values of which can be determined by the experimental conditions.

If the bar is considered to be infinitely *long*, its cold end will be nearly at the temperature of the surroundings, *i.e.*

as $x \to \infty, \theta \to 0.$

Applying this condition in eq. (i), we get

$$0 = a_1e^{\mu\infty} + a_2e^{-\mu\infty}$$

$$= a_1e^{\mu\infty} \qquad \left(\because -\mu_1e^{\mu\infty} = 0\right)$$

$$\therefore \quad a_1 = 0$$

Again, if θ_0 is the excess-temperature above the surroundings at the hot end $(x = 0)$ of the bar, we have

at $\quad x = 0, \quad \theta = \theta_0.$

Applying this condition and also putting $a_1 = 0$ in eq. (i), we have

$$\theta_0 = 0 + a_2e^{0.}$$

$$\therefore \quad a_2 = \theta_0 \qquad \left[\because e^0 = 1\right]$$

Putting the values of a_1 and a_2 in eq. (i), we get

$$\theta = \theta_0 e^{-\mu x} \qquad ...(ii)$$

This equation gives the temperature-distribution along each bar

Let θ_m be the melting point of wax (measured with respect to the surrounding temperature). Then $\theta = \theta_m$ at $x = l_1$ for bar 1, at $x = l_2$ for bar 2, at $x = l_3$ for bar 3.......... . Therefore, from eq. (ii), we have for 1,2, 3, bars

$$\theta_m = \theta_0 e^{-\mu_1 l_1}$$
$$\theta_m = \theta_0 e^{-\mu_2 l_2}$$
$$\theta_m = \theta_0 e^{-\mu_3 l_3}$$
$$.....................$$
$$.....................$$
$$\mu_1 l_1 = \mu_2 l_2 = \mu_3 l_3 = ...$$
$$\therefore \ \mu_1^2 l_1^2 = \mu_2^2 l_2^2 = \mu_3^2 l_3^2 = ...$$

But $\mu_1^2 = \dfrac{Ep}{K_1 A}, \mu_2^2 = \dfrac{Ep}{K_2 A}.........p, A$ and E being the same for all the bars since they are exactly identical in size, shape and polish.

$$\therefore \qquad \frac{Ep}{K_1 A} l_1^2 = \frac{Ep}{K_2 A} l_2^2 = \frac{Ep}{K_3 A} l_3^2 = ...$$

or
$$\frac{l_1^2}{K_1} = \frac{l_2^2}{K_2} = \frac{l_3^2}{K_3} = ...$$

or
$$\frac{K_1}{l_1^2} = \frac{K_2}{l_2^2} = \frac{K_3}{l_3^2} = ...$$

or
$$k_1 : k_2 : k_3 .... = l_1^2 : l_2^2 : l_3^2$$

Thus the conductivities of the different bars vary as the square of the length upto which wax is melted.

Forbes' Method for determining Conductivity : Forbes, in 1867, devised a method for determining the thermal conductivity K of a good conductor directly in terms of the definition of *K* as given by the relation

$$\frac{dQ}{dt} = -KA\frac{d\theta}{dx}$$

where $\frac{dQ}{dt}, A, \frac{d\theta}{dx}$ have their usual meanings.

Theory : Let us consider a long bar, one end of which is maintained at a constant high temperature. In the beginning the temperatures at various points along the bar gradually rise, but finally a steady state is reached at which these temperatures become stationary. In this state all the heat given to the bar escapes out from its surface into the surroundings.

Let us consider two cross-sections *L* and *M* of the bar. Let *A* be the area of cross-section of the bar, *K* the thermal conductivity of its material and $-\left\{\frac{d\theta}{dx}\right\}_L$ and $-\left\{\frac{d\theta}{dx}\right\}_M$ the temperature-gradients at *L* and *M* respectively. Then, the rate of heat-flow across the section *L* will be

$$-KA\left\{\frac{d\theta}{dx}\right\}_L$$

and that across the section *M* will be

$$-KA\left\{\frac{d\theta}{dx}\right\}_M$$

Therefore, the excess of heat flowing per second across *L* over that flowing across *M* is

$$-KA\left[\left(\frac{d\theta}{dx}\right)_L - \left(\frac{d\theta}{dx}\right)_M\right] \qquad \text{...(i)}$$

In the steady state, this amount of heat escapes out per

second from the surface of the bar between L and M. Now let us consider an element of length dx of the bar. If $-\frac{d\theta}{dt}$ be the rate of cooling of the element, c the specific heat and ρ the density of the material of the bar, then the heat escaped out per second from the element

$$= \text{mass x sp. heat x rate of cooling}$$

$$= (A\,dx\,\rho)\times c\times\left(-\frac{d\theta}{dt}\right)$$

Hence the total heat escaped out per second from all such elements between L and M is

$$= -\int_L^M A\rho c\frac{d\theta}{dt}dx = -A\rho c\int_L^M\frac{d\theta}{dt}dx \quad \text{...(ii)}$$

Equating (i) and (ii), we get

$$-KA\left[\left(\frac{d\theta}{dx}\right)_L - \left(\frac{d\theta}{dx}\right)_M\right] = -A\rho c\int_L^M\frac{d\theta}{dt}dx$$

$$\therefore \qquad K = \frac{\rho c\int_L^M\frac{d\theta}{dt}dx}{\left(\frac{d\theta}{dx}\right)_L - \left(\frac{d\theta}{dx}\right)_M} \quad \text{...(iii)}$$

Thus, in order to determine K, we have to obtain the values of

$$\left[\left(\frac{d\theta}{dx}\right)_L - \left(\frac{d\theta}{dx}\right)_M\right] \text{ and } \int_L^M\frac{d\theta}{dt}dx$$

Two separate experiments are performed to obtain the values of these quantities; a *statical* experiment for the former and a *dynamical* experiment for the latter.

Statical Experiment

The specimen metal is taken in the form of a long bar *AB*. It is held horizontal and its one end is immersed in a constant-temperature lead-bath. An insulating screen *S* is placed to shield the bar from direct radiation from the bath. A number of equidistant holes have been drilled all along

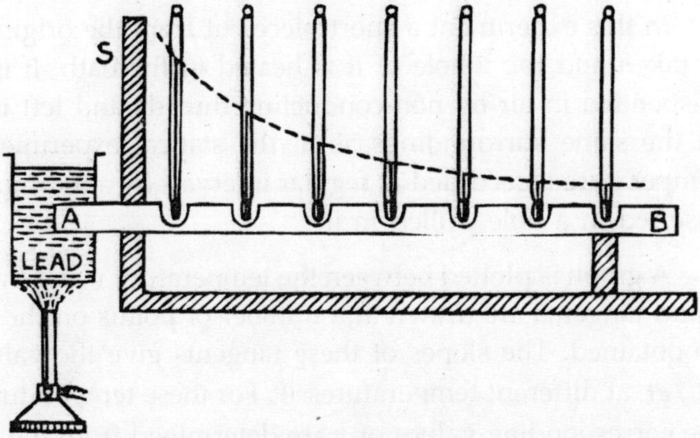

the bar. Sensitive thermometers are inserted in these holes through mercury contact. The arrangement is left for several hours until the steady state is reached.

A graph is plotted between the steady temperature θ and the distance x from the hot end. Tangents are drawn at two points *L* and *M* on this curve and the angles ϕ and ϕ_m, which they make with the x-axis, are measured. We have

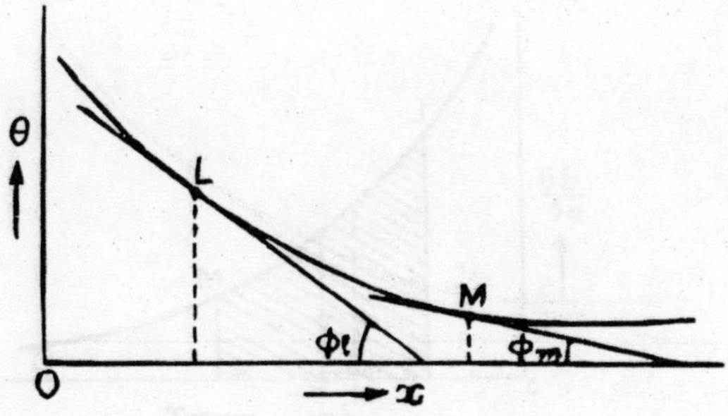

$$\left(\frac{d\theta}{dx}\right)_L = \tan\phi_l \text{ and } \left(\frac{d\theta}{dx}\right)_M = \tan\phi_m.$$

$$\therefore \qquad \left(\frac{d\theta}{dx}\right)_L - \left(\frac{d\theta}{dx}\right)_M = \tan\phi_l - \tan\phi_m.$$

which is thus obtained.

Dynamical Experiment

In this experiment a short piece cut from the original bar is taken and the whole of it is heated in the bath. It is then suspended in air by non-conducting threads and left to cool in the same surroundings as in the statical experiment. Its temperature is recorded at regular intervals by a thermometer inserted in a hole drilled in it.

A graph is plotted between the temperature θ and the time t and tangents are drawn at a number of points on the curve so obtained. The slopes of these tangents give the values of $d\theta/dt$ at different temperatures θ. For these temperatures θ, the corresponding values of x are determined from the graph between θ and x. Thus, we obtain a set of values of $d\theta/dt$ for different values of x.

A final graph is now plotted between $d\theta/dt$ and x. Two points L and M are taken on this curve for the same values of x as the points Land M on the curve between θ and x. Ordinates are drawn from L and M. If we consider a small strip of width dx as shown, its area will be $(d\theta/dt)dx$. Thus the quantity

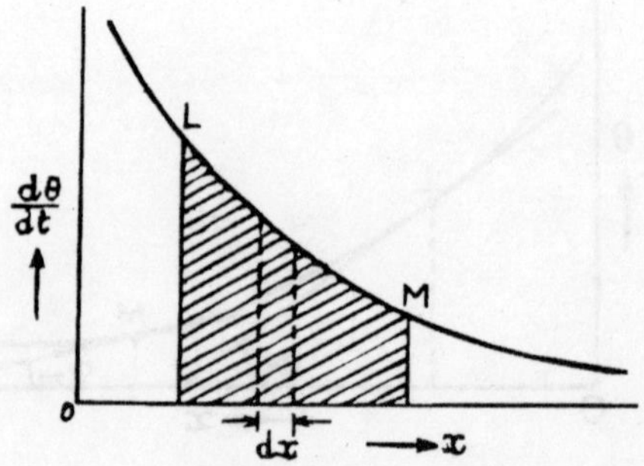

$$\int_L^M \frac{d\theta}{dt} dx$$

will represent the total shaded area between the curve and the ordinates from L and M. This can be measured by a planimeter, or can be calculated by drawing the curve on a graph paper.

We have thus obtained the value of $\left(\frac{d\theta}{dx}\right)_L - \left(\frac{d\theta}{dx}\right)_M$, as well as of $\int_L^M \frac{d\theta}{dt} dx$. The values of p and c can be taken from tables. Hence the conductivity K can be calculated from eq. (iii).

Drawbacks of the Experiment ;

(i) It takes 8-10 hours for the statical experiment,

(ii) The whole length of the bar does not attain a true steady state,

(iii) The temperature of the lead-bath does not remain strictly constant,

(iv) The distributions of temperature along the bar in the statical and dynamical experiments are different,

(v) The Newton's law of cooling is supposed to be true for large temperature-excesses which is hardly justified,

(vi) The specific heat of the bar does not remain constant at different temperatures as assumed.

This method cannot be used to determine the conductivity of a glass rod. Glass is a poor conductor of heat. Hence even at a large temperature-difference between the ends of the glass rod, the rate of flow of heat through it will be too small to be measured accurately.

Lees' Disc Method for Poor Conductors : An accurate method of measuring the thermal conductivity of poor conductors has been devised by Lees (and also by Charlton). The specimen, which is taken in the form of a flat disc M is placed between two copper discs C_2 and C_3, each about 4 cm in diameter and 3 mm thick. A little glycerine is applied to the faces of C_2 and C_T, to obtain good thermal contact between them and the faces of *M*. A thinner copper disc C_1 (about 1 mm thick) is placed on C_2, and a well-insulated heating coil *H* of platinoid wire is sandwiched between C_1 and C_2.

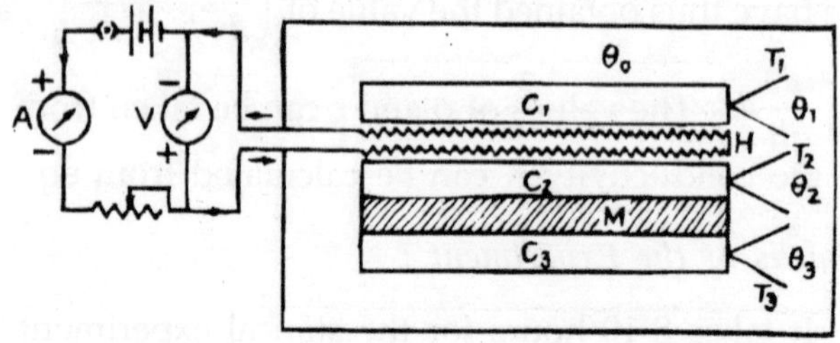

To measure the temperatures of C_1, C_2 and C_3, thermocouples T_1 T_2 and T_3 are inserted in tiny holes drilled in the edges of these discs. As copper is a very good conductor of heat, the readings of T_2 and T_3 give also the temperatures of the upper and the lower faces of *M*. The whole of this assembly is varnished on the outside to obtain a constant *emissivity* over the whole surface. It is then suspended by non-conducting threads in an enclosure maintained at a constant temperature.

Procedure : Before starting the actual experiment, the thermocouples are calibrated and their calibration curves are drawn.

The current in the heating-coil is switched on. The apparatus is left until a steady state is reached when the thermocouples show stationary temperatures. These temperatures are recorded. The readings of the ammeter A and the voltmeter *V* inserted in the circuit of the heating-coil *H* are taken.

Calculation : Let θ_1, θ_2 and θ_3 be the steady temperatures of the discs C_1, C_2 and C_3 respectively and θ_0 the temperature of the surrounding enclosure. The temperature of *M* may be taken as the mean of the temperatures of C_2 and C_3 *i.e.*, $\frac{\theta_2+\theta_3}{2}$.

In the steady state, a part of the heat entering the specimen M escapes out from its *exposed* surface and the remaining flows to C_3 and subsequently escapes from its *exposed* surface. If S_1, S_2, S_3 and S_M be the areas of the *exposed* surfaces of C_1, C_2, C_3 and M respectively and E the surface emissivity, then the heat escaped per second from M

= surface emissivity x surface area x temp-excess

$$= E \times S_M \times \left(\frac{\theta_2+\theta_3}{2} - \theta_0\right)$$

Similarly, the heat escaping per second from C_3

$$= E \times S_3 \times (\theta_3 - \theta_0)$$

Hence the heat entering the specimen per second

$$= ES_M\left(\frac{\theta_2+\theta_3}{2} - \theta_0\right) + ES_3(\theta_3 - \theta_0)$$

and the heat leaving the specimen per second

= heat lost per second from C_3

$$= ES_3(\theta_3 - \theta_0)$$

Therefore, the heat Q conducted through M per second, which may be regarded as the mean of that entering and leaving its faces, will be given by

$$Q = \frac{1}{2}\left[ES_M\left(\frac{\theta_2+\theta_3}{2} - \theta_0\right) + ES_3(\theta_3 - \theta_0) + ES_3(\theta_3 - \theta_0)\right]$$

$$= E\left[\frac{S_M}{2}\left\{\frac{\theta_2+\theta_3}{2} - \theta_0\right\} + S_3(\theta_3 - \theta_0)\right]$$

But if K be the thermal conductivity of the specimen, A its area of cross-section and d its thickness, then the heat flowing through it per second is

$$Q = KA\frac{\theta_2 - \theta_3}{d}$$

$$\therefore \quad KA\frac{\theta_2 - \theta_3}{d} = E\left[\frac{S_M}{2}\left\{\frac{\theta_2 + \theta_3}{2} - \theta_0\right\} + S_3(\theta_3 - \theta_0)\right] \qquad \ldots\text{(i)}$$

In the above expression, E is not known. We note that, in the steady state, the total heat dissipated per second in the heating-coil H is equal to the heat escaped per second from the *exposed* surface of the whole assembly. Hence if i (amp) be the current in the coil and V (volt) the p.d. across it, we have

$$\frac{Vi}{J} = E\left[S_1(\theta_1 + \theta_0) + S_2(\theta_2 - \theta_0) + S_M\left\{\frac{\theta_2 + \theta_3}{2} - \theta_0\right\} + S_3(\theta_3 - \theta_0)\right] \ldots\text{(ii)}$$

where J is number of joules per calorie. Hence K can be calculated from eq. (i) and (ii). Corrections are made for the small loss of heat from the edges of the heating-coil and the loss along the leads of the heating-coil.

This method cannot be used for good conductors because the temperature-difference between the two faces of a good-conductor disc will be too small to be measured.

Radial Heat-flow through a Cylindrical Shell : Let us consider a cylindrical shell of length l, and internal and external radii r_1 and r_2 respectively. Let its inner surface be maintained at a steady temperature θ_1 and the outer surface at a steady, but lower, temperature θ_2. Heat will flow from the inner side towards the outer side along the radii of the cylinder.

Let us consider an elementary cylindrical shell of radii r and $r + dr$. Let θ and $\theta + d\theta$ be the temperatures at r and $r + dr$. The temperature-gradient at r is - $(d\theta/dr)$. Since the shell is very thin and heat flows normal to the faces, it can be considered as a thin , parallel-faced slab. Hence the rate of heat-flow through it is given by

$$\frac{dQ}{dt} = -K(2\pi rl)\frac{d\theta}{dr}$$

where K is the thermal conductivity and $2\pi rl$ the surface area of the shell. In the steady state, dQ/dt is same for all values of r. Re-arranging, we get

$$d\theta = \frac{(dQ/dt)}{2\pi Kl}\frac{dr}{r}.$$

Integrating, we get

$$\theta = -\frac{(dQ/dt)}{2\pi Kl}\log_e r + C(a \text{ constant}) \quad .(i)$$

Now, when $r = r_1$, $\theta = \theta_1$.

$$\therefore \quad \theta_2 = -\frac{(dQ/dt)}{2\pi Kl}\log_e r_1 + C \quad ...(ii)$$

When r = r_2, $\theta = \theta_2$.

$$\therefore \quad \theta_2 = -\frac{(dQ/dt)}{2\pi Kl}\log_e r_2 + C \quad ...(iii)$$

Subtracting eq. (iii) from eq. (ii), we get

$$(\theta_1 = \theta_2) = \frac{(dQ/dt)}{2\pi Kl}\log_e \frac{r_2}{r_1}$$

from which

$$\frac{dQ}{dt} = 2\pi Kl\frac{(\theta_1 - \theta_2)}{\log_e (r_2/r_1)} \quad ...(iv)$$

This expression can be used to determine the thermal conductivity of poor conductors given in the form of cylindrical tubes.

Temperature Distribution

In order to get expression for the temperature distribution, let us substitute the value of dQ/dt from eq: (iv) in *eq.* (ii):

$$\theta_1 = \frac{\theta_1 - \theta_2}{\log_e (r_2/r_1)} \log_e r_1 + C$$

$$\therefore \quad C = \theta_1 + \frac{(\theta_1 - \theta_2)\log_e r_1}{\log_e (r_2/r_1)} \quad \ldots(v)$$

Substituting the values of dQ/dt and C from eq. (iv) and (v) in eq (i), we get

$$\theta = -\frac{(\theta_1 - \theta_2)}{\log_e (r_2/r_1)} \log_e r + \theta_1 + \frac{(\theta_1 - \theta_2)\log_e r_1}{\log_e (r_2/r_1)}$$

$$\text{or} \quad \theta = \theta_1 - (\theta_1 - \theta_2)\frac{\log_e (r/r_1)}{\log_e (r_2/r_1)}$$

This is the required expression.

Determination of Thermal Conductivity of Rubber Tube (Cylindrical Shell Method) : A measured length l of the given rubber tube is immersed in a single loop in a known amount of water taken in a calorimeter C. The initial temperature of water is recorded by a thermometer *T*. Now; by opening the stop-cock S_1, the steam from a boiler *B* is passed through the specimen tube, whose inner surface attains the temperature of steam. Heat flows radially from the inside to the outside of the tube and raises the temperature of the water in the calorimeter. When the temperature of water rises by about 15° C, the steam is cut off. The time t for which the steam has been passed is noted.

Let θ' and θ'' be the initial and final temperatures of water. If m be the mass of the water (specific heat 1) and w the water equivalent of the calorimeter, the quantity of heat received by the calorimeter and its contents per second is given by

$$Q = \frac{(m + w)(\theta'' - \theta')}{t} \quad \ldots(vi)$$

This must be the heat flown radially per second through the walls of the tube immersed in water *i.e.*

$$Q = 2\pi K l \frac{(\theta_1 - \theta_2)}{\log_e \frac{r_2}{r_1}} = 2\pi K l \frac{(\theta_1 - \theta_2)}{\log_e \frac{b}{a}}$$

where a and b are the internal and external diameters, and θ_1 and θ_2 the temperature of the inner and outer surfaces respectively of the tube.

Here $\theta_1 = \theta_{\text{stream}}$ (temperature of steam) and $\theta_2 = \frac{\theta' + \theta''}{2}$ (the temperature of the outer surface may be taken as the mean of the initial and final temperatures of water). Hence the last equation will be written as

$$Q = 2\pi K l \frac{\theta_{\text{stream}} - \frac{\theta' \div \theta''}{2}}{\log_e (b/a)} \qquad \text{...(vii)}$$

Eq. (vi) and (vii) give

$$2\pi K l \frac{\theta_{\text{stream}} - \frac{\theta' + \theta''}{2}}{\log_e \frac{b}{a}} = \frac{(m + w)(\theta'' - \theta')}{t}$$

$$\therefore \qquad K = \frac{(m + w)(\theta'' - \theta') \log_e (b/a)}{2\pi l \left[\theta_{\text{steam}} - \frac{\theta' + \theta''}{2}\right] t}$$

From this the thermal conductivity of rubber, K, is calculated.

Thermal Conductivity of a Glass Tube : The thermal conductivity of a thick-walled glass tube can be determined by allowing heat to flow radially through the walls of the tube, as indicated.

The glass tube is placed in a slightly tilted position and a slow stream of cold water is passed through it from a constant-head device. Two $\frac{1}{10}^\circ$ thermometers T_1 and T_2 are inserted in glass-connecting pieces attached to the ends of the tube as shown. The tube is surrounded by a jacket through which steam circulates. The outer surface of the tube attains the temperature of steam and the heat flows radially through its wall towards the inside. As a result, the temperature of water flowing through the tube from one end to another rises.

When the steady state is attained, the temperatures of the in flowing and out flowing water are recorded and the mass of water flowing through the tube in a known interval of time is measured. The length l of the part of the tube surrounded by steam, and its internal and external diameters are measured.

If θ' and θ'' be the temperatures of the in flowing and out flowing water and m its mass flowing in time t, the quantity of heat received by water (specific heat 1) per second is

$$Q = \frac{m(\theta''-\theta')}{t} \qquad \text{...(i)}$$

This is also the heat flowing per second through the walls of the tube from the outside to the inside.

Let θ_2 and θ_1 be the temperatures of the outer and inner surfaces of the glass tube, and r_2 and r_1 its external and internal radii respectively. Let us consider a thin cylindrical shell of radii r and $r + dr$. Let θ and $\theta + d\theta$ be the temperatures at distances r and $r + dr$ respectively from the centre. The temperature-gradient at r is $+ (d\theta/dr)$. Considering the shell as a thin parallel-faced slab, the quantity of heat flowing per second through it is given by

$$Q = K(2\pi rl)\frac{d\theta}{dr}$$

where K is the thermal conductivity and $2\pi rl$ the surface area of the shell. Rearranging, we get

$$\frac{dr}{r} = \frac{2\pi Kl}{Q} d\theta$$

Integrating, we get

$$\log_e r = \frac{2\pi Kl}{Q}\theta + C(\text{a constant}),$$

Now , when $r = r_1$, $\theta = \theta_1$

$$\therefore \quad \log_e r_1 = \frac{2\pi Kl}{Q}\theta_1 + C \qquad \text{...(ii)}$$

and when $r = r_2$, $\theta = \theta_2$,

$$\therefore \quad \log_e r_2 = \frac{2\pi Kl}{Q}\theta_2 + C \qquad \text{...(iii)}$$

Subtracting eq. (ii) from eq. (iii), we get

$$\log_e \frac{r_2}{r_1} = \frac{2\pi Kl}{Q}(\theta_2 - \theta_1)$$

from which $$Q = 2\pi Kl \frac{(\theta_2 - \theta_1)}{\log_e \frac{r_2}{r_1}}$$

Now, the temperature of the outer surface of the glass is equal to the temperature of the steam *i.e.* $\theta_2 = \theta_{steam}$. The temperature of the inner surface may be taken as the mean of the temperatures of the in flowing and out flowing water *i.e.* $\theta_1 = \frac{\theta' + \theta''}{2}$. Further, if a and b be the internal and external diameters of the tube, then $r_1 = a/2$ and $r_2 = b/2$. Making these substitutions, the last expression becomes

$$Q = 2\pi Kl \frac{\left(\theta_{steam} - \frac{\theta' + \theta''}{2}\right)}{\log_e \frac{b}{a}} \qquad \text{...(iv)}$$

This is the quantity of heat flowing per second through the walls of the tube.

Now, eq. (i) and (iv) give

$$2\pi Kl\frac{\left(\theta_{steam}-\frac{\theta'+\theta''}{2}\right)}{\log_e\frac{b}{a}}=\frac{m(\theta''-\theta')}{t}$$

$$\therefore \qquad K=\frac{m(\theta''-\theta')\log_e\frac{b}{a}}{2\pi l\left(\theta_{steam}-\frac{\theta'+\theta''}{2}\right)t}$$

Hence K can be calculated.

Radial Heat Flow through a Spherical Shell: Let O be the centre of a spherical shell and A and B its inner and outer walls respectively. Let r_1 and r_2 be the radii, and θ_1 and θ_2 the steady temperatures of the inner and outer walls. Let us consider an elementary spherical shell of radii r and $r + dr$. Let θ and $\theta +$ $d\theta$ be the temperatures at distances r and $r + dr$ respectively from O. The temperature-gradient at r is $-(d\theta/dr)$. Then the rate of heat-flow across the surface of the shell of radius r is

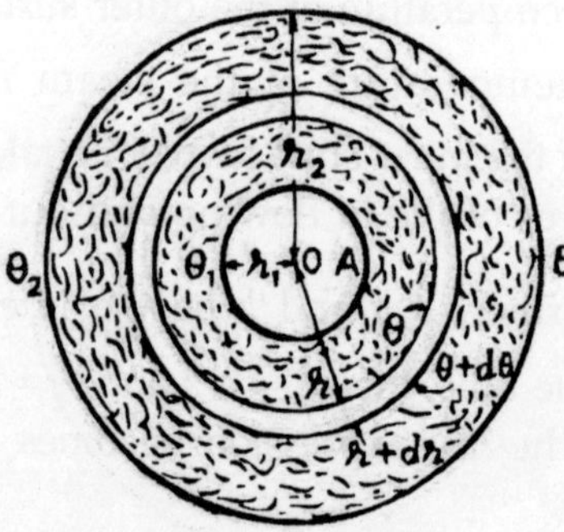

$$\frac{dQ}{dt}=-K\left(4\pi r^2\right)\frac{d\theta}{dr},$$

where K is the thermal conductivity of the specimen. Due to symmetry, the heat flows out equally in all directions in the sphere and in the steady state, dQ/dt is constant for all values of r. Hence, rearranging the above expression, we get

$$d\theta = \frac{(dQ/dt)}{4\pi K}\frac{dr}{r^2}.$$

Integrating: $$\theta = \frac{(dQ/dt)}{4\pi K}\frac{1}{r} + C, \qquad \text{...(i)}$$

where C is integration constant.

Now, when $r = r_1$, $\theta = \theta_1$.

$$\therefore \quad \theta_1 = \frac{(dQ/dt)}{4\pi K}\frac{1}{r_1} + C \qquad \text{...(ii)}$$

when $r = r_2$, $\theta = \theta_2$.

$$\therefore \quad \theta_2 = \frac{(dQ/dt)}{4\pi K}\frac{1}{r_2} + C \qquad \text{...(iii)}$$

Subtracting eq. (iii) from (ii), we get

$$(\theta_1 - \theta_2) = \frac{(dQ/dt)}{4\pi K}\left(\frac{1}{r_1} - \frac{1}{r_2}\right)$$

or $$\frac{dQ}{dt} = \frac{4\pi K(\theta_1 - \theta_2)}{\left[\frac{1}{r_1} - \frac{1}{r_2}\right]} = \frac{4\pi K r_1 r_2 (\theta_1 - \theta_2)}{r_2 - r_1} \qquad \text{...(iv)}$$

Temperature Distribution : Substituting the value of dQ/dr from eq. (iv) in eq. (ii), we get

$$\theta_1 = \frac{\theta_1 - \theta_2}{\frac{1}{r_1} - \frac{1}{r_2}}\left[\frac{1}{r_1}\right] + C$$

$$\therefore \quad C = \theta_1 - \frac{\theta_1 - \theta_2}{\frac{1}{r_1} - \frac{1}{r_2}}\left[\frac{1}{r_1}\right] \qquad \ldots(v)$$

Substituting the values of dQ/dt and C from eq. (iv) and (v) in eq. (i), we get

$$\theta = \frac{\theta_1 - \theta_2}{\frac{1}{r_1} - \frac{1}{r_2}}\left[\frac{1}{r_1}\right] + \theta_1 - \frac{\theta_1 - \theta_2}{\frac{1}{r_1} - \frac{1}{r_2}}\left[\frac{1}{r_1}\right]$$

$$= \theta_1 + (\theta_1 - \theta_2)\frac{\frac{1}{r} - \frac{1}{r_1}}{\frac{1}{r_1} - \frac{1}{r_2}} \qquad \ldots(vi)$$

Determination of Thermal Conductivity (Spherical Shell Method) : Two thin copper spherical shells A and B, which can be split into hemispheres are selected. The specimen under test is filled in the space between A and B . A source of constant heat-supply, *e.g.*, an electric heating-element, is placed at the centre O of the shells. The heat is conducted radially through the specimen and subsequently lost to the surroundings from the surface of the outer shell. The temperatures of the two shells, when steady state is reached, are recorded by thermocouples inserted in them. If i amp be the current and V volt the potential difference across the heating-element, then the heat supplied per second at the centre of the shells is

$$Q = \frac{Vi}{J} \text{cal},$$

where J is the number of joules/calorie.

But, in the steady state, the heat flowing out radially per second is, by eq. (iv),

$$Q = \frac{4\pi K r_1 r_2 (\theta_1 - \theta_2)}{r_2 - r_1}$$

Then $$\frac{4\pi K r_1 r_2 (\theta_1 - \theta_2)}{r_2 - r_1} = \frac{Vi}{J}$$

$$\therefore \qquad K = \frac{(r_2 - r_1)Vi}{4\pi(\theta_1 - \theta_2) r_1 r_2 J}$$

Hence K may be calculated.

Periodic Flow of Heat

When one end of a bar is alternately heated and cooled, the heat flows through the bar periodically. Then , the temperature at any given point on the bar also varies periodically.

Let us assume that the temperature at the hot end of a bar, insulated from surroundings, varies simple harmonically, and is given by

$$\theta = \theta_0 \cos \omega t, \qquad \text{...(i)}$$

where θ_0 is the temperature-amplitude. Let the direction of heat-flow be along the x-axis. Then the Fourier's differential equation will be

$$\frac{d\theta}{dt} = k\frac{d^2\theta}{dx^2} \qquad \text{...(ii)}$$

where k is the diffusivity ($k = \frac{k}{\rho c}$ where K is conductivity, ρ density and c is sp. heat).

To solve this equation, let us try a solution

$$\theta = Ae^{\alpha x + i\beta t} \qquad \text{...(iii)}$$

where A, α, β are constants and $i = \sqrt{-1}$. This gives

$$\frac{d\theta}{dt} = Ae^{\alpha x + i\beta t}(i\beta) \text{ and } \frac{d^2\theta}{dx^2} = Ae^{\alpha x + i\beta t}(\alpha^2).$$

Substituting in eq. (ii), we get

$$i\beta = k\alpha^2$$

$$\therefore \quad \alpha = \pm\sqrt{\frac{i\beta}{k}}$$

Now, $(1 + t)^2 = 2i$ and so $\sqrt{i} = \frac{1}{\sqrt{2}}(1+i)$. Hence

$$\alpha = \pm\sqrt{\frac{\beta}{2k}}(1+i)$$

Substituting this value in eq. (iii), we get

$$\theta = Ae^{\pm\sqrt{\beta/2k}(1+i)x+i\beta t}.$$

When $x = \infty$ (at cold end), $\theta \neq \infty$. Hence only the negative sign is admissible. Separating the real and imaginary parts, we get

$$\theta = Ae^{-\sqrt{\beta/2k}x+i\left(\beta t-\sqrt{\beta/2k}x\right)}$$

$$= Ae^{-\sqrt{\beta/2k}\,x} + \left[\cos\left\{\beta t - \sqrt{\frac{\beta}{2k}}x\right\} + i\sin\left\{\beta t - \sqrt{\frac{\beta}{2k}}x\right\}\right]$$

Rejecting the imaginary term, we have

$$\theta = Ae^{-\sqrt{\beta/2k}\,x}\cos\left[\beta t - \sqrt{\frac{\beta}{2k}}x\right]$$

At $x = 0$ (hot end), this gives $\theta = A\cos\beta t$. Comparing this with eq. (i), we have $A = \theta_0$ and $\beta = \omega$.

Hence we have

$$\theta = \theta_0 Ae^{-\sqrt{\omega/2k}}\cos\left(\omega t - \sqrt{\frac{\omega}{2k}}x\right)$$

This represents a heat wave travelling with velocity

$\sqrt{2\omega k} = \sqrt{\frac{4\pi k}{T}}$. It shows that at a particular point along the bar (x = constant), the temperature θ varies harmonically with time, the period being that of the heat source. It further shows that the amplitude of temperature oscillations diminishes exponentially as the distance x along the bar increases and becomes negligible at a sufficient distance.

Angstrom's Method

Angstrom used the method of periodic flow of heat to determine the thermal conductivity of a metal bar. He enclosed one end of the bar in a chamber through which steam and water could be alternately passed. The bar was heated for 12 minutes and cooled also for the same time. Thus the period of heat flow was T = 24 minutes. Temperatures were observed every minute at two points on the bar by means of two thermocouples. The diffusivity was calculated from the formula

$$k = \frac{n\pi l^2}{T(\delta_n - \delta_n')\log_e(\beta_n/\beta_n')},$$

where l is the distance between the two points and δ_n, δ_n', β_n, β_n' are constants for the two points which were evaluated graphically from the observations.

Conductivity of Earth's Crust

The method of periodic heat flow is very suitable for finding the thermal conductivity of earth's crust. At any point the earth's surface is heated in the day and cooled in the night. This alternate heating and cooling travels into the interior of the earth in the form of a heat wave with a period, T= 24 hours. Again, at any point the earth receives larger heat in the summer than in winter, so that a second heat wave with a period, T= 1 year is propagated into the interior of the earth.

Forbes embedded a number of thermometers at different

depths in the earth and investigated the progress of the heat wave inside the earth. The time t in which a temperature at the surface reaches a point at a depth d is

$$t = \frac{d}{v} = \sqrt{\frac{T}{4\pi k}}\, d$$

By measuring t for various values of d, he calculated the diffusivity of earth's crust at several places. Such investigations have also been used in determining the penetration of the daily and annual changes of temperature within the earth's crust. The information is very useful for geologists.

Kelvin calculated the time taken by the earth to cool from its initial molten fluid state to its present temperature, and estimated a value of 100 million years for the age of the earth. His estimate, however, falls short of the value given by geologists. This is due to the presence of radioactive substances inside the earth which continuously generate heat by spontaneous disintegration, and retard the cooling of the earth.

It is a common experience that in summer, while the outside temperature is maximum in mid-day, the temperature inside a closed room reaches its maximum value late in the evening. The reason is that the diurnal heat-wave of period T= 24 hours (due to alternate sun-heating and cooling of earth) propagates very slowly. The roofs and walls of a room are exposed to the heat of the sun. Taking k = 0.006 cgs units for concrete and d = 24 cm (thickness of walls), the time taken by the heat-wave to penetrate into the room is

$$t = \sqrt{\frac{T}{4\pi k}}\, d = \sqrt{\frac{24 \times 60 \times 60}{4 \times 3.14 \times 0.006}} \times 24$$

$$= 6 \text{ hours.}$$

Thus, if the temperature outside the room becomes maximum at 2 PM, the inside of the room will be hottest at about 8 PM when the walls begin to radiate. Hence it becomes highly uncomfortable to sleep in the room at this time.

Relation between Thermal and Electrical Conductivities of Metals : (Wiedemann-Franz Law) : It is a common observation that thermal and electrical conductivities of metals go together. All good conductors of heat are also good conductors of electricity. This was expressed by Wiedemann and Franz in the form of an empirical law according to which *"the ratio of the thermal and electrical conductivities is the same for all metals at the same temperature" i.e.*

$$\frac{K}{\sigma} = \text{constant}$$

Lorentz extended the law and showed that *this ratio is proportional to the absolute temperature i.e.*

$$\frac{K}{\sigma T} = \text{constant.}$$

Deduction: Drude gave a theoretical proof of this law. He assumed that in a metal, there are a large number of "free" electrons which are in a state of random motion, like the molecules of a gas. These free electrons are responsible for both kinds of conductivity. The average drift of these electrons in any direction under the influence of a difference of potential constitute the electric current; and the transfer of energy of random motion in any direction is thermal conduction.

Let us consider the conduction of electricity in a metallic wire across which a potential difference has been applied. Let the potential difference create an electric intensity X along the wire. Then every free electron in the wire is acted upon by a force Xe and has an acceleration Xe/m, where e is the charge and m the mass of the electron. As the electrons behave like gas molecules, they collide and are in equilibrium with the atoms of the metal.

Let λ be the mean free path (mean distance travelled between two successive collisions) and $\bar{c}$ the mean velocity of the electrons at the temperature T. The electron is accelerated between two collisions, but loses its velocity on collision with the atom. Thus its velocity at the beginning of the path is zero and at the end = acceleration x time = $\frac{Xe}{m} \times \frac{\lambda}{c}$. Hence the 'average' drift velocity of the electron is given by

$$u = \frac{1}{2}\left\{0 + \frac{Xe\lambda}{m\bar{c}}\right\} = \frac{Xe\lambda}{2m\bar{c}}$$

If n be the number of free electrons per unit volume, the number crossing unit cross-sectional area per second would be nu. Then, the current flowing per unit area is

i = number of electrons flowing per unit area per second × charge on each electron

$$= nu \times e$$

$$= n \times \frac{Xe\lambda}{2m\bar{c}} \times e = \frac{ne^2\lambda X}{2m\bar{c}}$$

The electrical conductivity is given by

$$\sigma = \frac{i}{X} = \frac{ne^2\lambda}{2m\bar{c}}$$

But, from kinetic theory, $\frac{1}{2}m\bar{c}^2 = \frac{3}{2}kT$, or $m\bar{c} = \frac{3kT}{\bar{c}}$, where k is the Boltzmann constant.

$$\therefore \quad \sigma = \frac{ne^2\lambda\bar{c}}{6KT} \qquad \text{...(i)}$$

Now, the thermal conductivity, K, for the electron-gas is given by

$$K = \frac{1}{3}mn\bar{c}(c_v)\lambda \qquad \text{...(ii)}$$

$m\,c_v$ is the thermal capacity of a single travelling particle. The increase in its energy due to increase in temperature by dT is

$$dE = m\,c_v\,dT$$

or $$mc_v = \frac{dE}{dT}$$

But $E = \frac{3}{2}kT$ (if the electron is assumed to possess only the translational motion).

$$\therefore \quad mc_v = \frac{3}{2}k$$

Substituting this value of $m\,c_v$ in eq. (ii), we get

$$K = \frac{n\bar{c}\lambda k}{2} \qquad \text{...(iii)}$$

Dividing eq. (iii) by eq. (i), we get

$$\frac{K}{\sigma} = \frac{3k^2T}{e^2}$$

or $$\frac{K}{\sigma T} = 3\left\{\frac{k}{e}\right\}^2 = \text{constant.}$$

This is Wiedemann-Franz-Lorentz law. At ordinary temperatures, the values for $K/\sigma\,T$ are actually found to be larger than that given by the law. The value of $K/\sigma\,T$ falls with the fall in temperature. Both the thermal and electrical conductivities increase with fall in temperature, but the electrical one at a greater rate.

Objections against Drude Theory :

(i) It has been assumed that the free electrons are responsible for both conductivities. It is, however, found that even the best insulators (which have no free electrons) have some thermal conductivity. This means that electrons alone are not responsible for thermal conduction.

(ii) It has been found that the free electrons make little or no contribution to the specific heat of the metals. This goes against Drude's assumption that the electrons are the carriers of thermal energy, like gas molecules.

(iii) The electrical resistance is found to drop sharply at very low temperatures and practically vanishes in the neighbourhood of absolute zero (super-conductivity state). According to Drude's free-electron theory, the electronic mean-free path must become very large or infinite in this state, which seems inconceivable.

These difficulties were finally resolved by the development of quantum statistics.

PROBLEMS

1. The thickness of ice in a lake is 5 cm and the temperature of air is –10 °C. Calculate the time required for the thickness of ice to be doubled. Constants for ice are : conductivity = 0.004 cal-c m^{-1}-se c^{-1}-° C^{-1}, density = 0.92 g/cm³, latent heat = 80 cal/g.

Solution : The cold air (below 0 °C) above the water in a lake takes in heat from the water. The water therefore begins to freeze into ice layer. Let us determine the rate of growth of this layer.

Let us consider a layer of ice x cm thick already formed on a lake at 0°C, the temperature of air above it being $-\theta$ °C. Let A be the area of the ice layer, L the latent heat of fusion of ice and ρ its density. Then the heat given up when the ice layer increases in thickness by an amount dx

= mass x latent heat

= $A\ dx\ \rho \times L$ calorie.

If this quantity of heat is conducted upwards through the ice layer in *dt* second, then we have

$$A\,dx\,\rho \times L = KA\frac{0-(-\theta)}{x}dt,$$

where *K* is thermal conductivity of ice. This gives

$$\frac{dx}{dt} = \frac{K\theta}{\rho L x}$$

dx/dt is the rate at which the thickness of ice layer increases. Rearranging the above expression, we get

$$dt = \frac{\rho L x}{K\theta}dx$$

Therefore, the time in which thickness of ice will increase from x_1 to x_2 will be given by

$$t = \frac{\rho L}{K\theta}\int_{x_1}^{x_2} x\,dx$$

$$= \frac{\rho L}{K\theta}\left[\frac{x^2}{2}\right]_{x_1}^{x_2}$$

$$= \frac{\rho L}{2K\theta}\left(x_2^2 - x_1^2\right)$$

In the given problem, we have

$K = 0.004$ cal-cm^{-1} -sec^{-1} -°C^{-1}, $\rho = 0.92$ g/cm^3,

$L = 80$ cal/g , $-\theta = -10$ °C, $x_1 = 5$ cm, $x_2 = 10$ cm (doubled).

$$\therefore \quad t = \frac{0.92 \times 80}{2 \times 0.004 \times 10}\left(10^2 - 5^2\right)$$

$$= 69000 \text{ seconds} = 19.1 \text{ hours.}$$

2. *How much time will it take for a layer of ice of thickness 20 cm to increase by 10 cm on the surface of a pond when the temperature of the surroundings is –15 °C. Given : K = 0.005 cgs units, L = 80 cal/g, ρ = 0.90 g/cm³.*

Solution : As in the last problem :

$$t = \frac{\rho L}{2K\theta}\left(x_2^2 - x_1^2\right)$$

$$= \frac{0.90 \times 80}{2 \times 0.005 \times 15}\left(30^2 - 20^2\right)$$

$$= \frac{0.90 \times 80 \times 500}{2 \times 0.005 \times 15}$$

$$= 240000 \text{ sec}$$

$$= 66.67 \text{ hours.}$$

3. A lake is covered with ice 4 cm thick and the temperature of air above the ice is –12 °C. At what rate, expressed in cm per hour, will the ice thicken ? Conductivity of ice = 0.0052 cgs units, density of ice = 0.92 g/cm³ and latent heat of ice = 80 cal/g.

Solution :

$$\frac{dx}{dt} = \frac{K\theta}{\rho L x}$$

$$= \frac{0.0052\,\text{cal}/(\text{cm-sec-°C}) \times (12\,°\text{C})}{0.92\,\text{g}/\text{cm}^3 \times 80\,\text{cal}/\text{g} \times 4\text{ cm}}$$

$$= 2.12 \times 10^4 \text{ cm/sec}$$

$$= (2.12 \times 10^{-4}) \times 3600$$

$$= 0.763 \text{ cm/hour.}$$

4. The steady temperatures at the ends of a copper rod of length 25 cm and area of cross-section 1.0 cm² are 125 °C and 0 °C respectively. Calculate the temperature-gradient, rate of heat-flow, and the temperature at a distance 10 cm from the hot end. K for copper = 9.2 × 10⁻² kilo-cal/(sec-meter-°C).

Solution : The rate of linear, steady heat-flow is given by

$$\frac{\Delta Q}{\Delta t} = KA\frac{\theta_1 - \theta_2}{d} \quad \text{...(i)}$$

where the symbols have their usual meanings.

Here, temperature-gradient, $\frac{\theta_1 - \theta_2}{d} = \frac{125}{25} = 5°C/cm$

$$K = 9.2 \times 10^{-2} \text{ kilo-cal/(sec-meter-°C)}$$

$$= 9.2 \times 10^{-1} \text{ cal/(sec-cm-°C)},$$

and $$A = 1.0 \text{ cm}^2.$$

$$\therefore \quad \frac{\Delta Q}{\Delta t} = (9.2 \times 10^{-1}) \times 1.0 \times 5$$

$$= 4.6 \text{ cal/sec.}$$

Now, in the steady state, the rate of heat-flow is same through all the sections of the bar. If the temperature at a distance of 10 cm from the hot end (*i.e.* d=*10* cm) be θ; then from eq. (i), we have

$$\frac{\Delta Q}{\Delta t} = 4.6 = 9.2 \times 10^{-1} \times 1.0 \times \frac{125 - \theta}{10}$$

or $$(125 - \theta) = \frac{4.6 \times 10}{9.2 \times 10^{-1}} = 50$$

$$\therefore \quad \theta = 125 - 50$$

$$= 75 \text{ °C.}$$

5. Three bars each of area of cross-section A and length L are connected in series. The thermal conductivities of their materials are K, 2K and 1.5 K. If the temperatures of the external ends of the first and the last bar are 200 °C and 18 °C, then calculate the temperatures of both the junctions. The loss of heat due to radiation is negligible.

Solution : Let the temperatures of the junctions be θ_1 and θ_2 respectively. The loss of heat due to radiation is negligible. Therefore, in the steady state, the rate of flow of heat in the whole system will be same. Therefore

$$\frac{\Delta Q}{\Delta t} = KA\frac{(200-\theta_1)}{L}$$

$$= (2K)A\frac{(\theta_1-\theta_2)}{L}$$

$$= (1.5K)A\frac{(\theta_2-18)}{L}$$

or $$200-\theta_1 = 2\theta_1-2\theta_2$$

$$= 1.5\theta_2-27$$

Solving : $$\theta_1 = 116°C,$$

$$\theta_2 = 174°C.$$

6. A bar of length 40 cm and uniform cross-section 5 cm² consists of two halves, AB of copper and CB of iron welded at B .The end A is maintained at 200 °C and C at 0 °C. The sides are thermally insulated. Find the rate of heat-flow along the bar in the steady state. K for Cu and Fe are 0.9 and 0.1 cal/(sec-cm-°C) respectively.

Solution : Let θ be the temperature of the junction. In the steady state, the rate of heat-flow will be same throughout the bar.

Therefore $$\frac{\Delta Q}{\Delta t} = 0.9\times5\times\frac{200-\theta}{20}$$

$$= 0.1\times5\times\frac{\theta-0}{20}$$

Solving for θ, we get

$$\theta = 180°$$

Now, putting the value of θ in the last eq., we get

$$\frac{\Delta Q}{\Delta t} = 0.1 \times 5 \times \frac{180 - 0}{20}$$
$$= 4.5 \text{ cal/sec.}$$

7. An electric hot plate of 100 cm² area has temperature of 820 °C on the heater side. A kettle of a perfect conductor with 250 cm³ of water at a temperature of 20 °C is kept over it. Find how much time will be required for the water to just boil if the water equivalent of the kettle is 20 gm and thickness of the hot plate is 0.9 cm and thermal conductivity is 0.9 C.G.S: units.

Solution : The mass of the water is 250 gm and the water equivalent of the kettle is 20 gm. The specific heat of water is 1C. G. S. unit. Hence the heat required for the water to reach from 20 °C to 100 °C (boiling) is given by

$$Q = (250 \times 1 + 20)(100 - 20)$$
$$= 21600 \text{ cal.}$$

Let t be the time required for the heat Q to flow through the hot plate. We know that, in usual notations,

$$Q = KA\frac{\theta_1 - \theta_2}{d}t$$

Here Q = 21600 cal, K = 0.9 C.G.S. units,

$$A = 100 \text{ cm}^2, (\theta_1 - \theta_2)$$
$$= \left\{820 - \frac{20 + 100}{2}\right\} = 160 \text{ °C}$$

and $$d = 0.9 \text{ cm}.$$

$$\therefore \quad 21600 = \frac{0.9 \times 100 \times 760 \times t}{0.9}$$

or $$t = \frac{21600 \times 0.9}{0.9 \times 100 \times 760} = \frac{27}{95} \text{sec.}$$

8. An insulating wall of area of 4 m^2 is made from cork, 16.5 cm thick, protected externally by 12.5 cm brick and lined on the inside with 10.0 cm wood. If the exposed surface of wood is maintained at – 5.0 °C and the outer surface of the brick at 20.0 °C, calculate (i) the rate of energy transfer through the wall and (ii) the interface temperatures. Given: thermal conductivity for brick = 2.5 × 10^{-3}, for cork = 1.1 × 10^{-4} and for wood = 4.0 x 10^{-4} cal-sec^{-1} - cm^{-1} - °C^{-1}.

Solution: Let θ_1 be the temperature of the brick-cork interface and θ_2 that of the cork-wood interface. The temperature of the outer surface of brick is 2.0 °C and that of the exposed surface of wood is –5 °C. In the steady state, the heat flowing per second is same throughout. Therefore, equating the heat flowing per second through brick, cork and wood, we have

$$2.5\times10^{-3}\times A\frac{20-\theta_1}{12.5} = 1.1\times10^{-4}\times A\frac{\theta_1-\theta_2}{16.5}$$

$$= 4.0\times10^{-4}\times A\frac{\theta_2-(-5)}{10}$$

From this , we get two equations

$$25\times\frac{20-\theta_1}{12.5} = 1.1\times\frac{\theta_1-\theta_2}{16.5}$$

and $$25\times\frac{20-\theta_1}{12.5} = 4\times\frac{\theta_2+5}{10}$$

Solving we get:

$$\theta_1 = 19.3°C \text{ and } \theta_2 = -1.53°C.$$

The rate of heat-flow through brick, say, is

$$\frac{\Delta Q}{\Delta t} = 2.5\times10^{-3}\times A\times\frac{20-\theta_1}{12.5}$$

$$= 2.5 \times 10^{-3} \times (400 \times 400) \times \frac{20 - 19.3}{12.5}$$

$$= 22.4 \text{ cal/sec.}$$

9. In Lees' experiment for finding the thermal conductivity of cardboard, the following readings were taken :

Steady temperature of copper disc = 99.5 °C. Steady temperature of lower disc = 83.5 °C . Time taken for the lower disc to cool from 86 °C to 81 °C = 4 min. Thickness of cardboard disc = 4.8 mm. Thickness of lower disc = 1 cm. The discs are of copper whose density is 9 gm/cm³ and sp heat is 0.1 cal/(g-°C). Find the conductivity of cardboard.

Solution: The heat conducted per second from the upper to the lower disc *through* the cardboard disc is

$$Q = K(\pi r^2)\frac{\theta_1 - \theta_2}{d}$$

where r is the radius and d the thickness of the cardboard disc. In the steady state, this heat is emitted away per second from the lower disc. Thus, if m is the mass, s the specific heat and $d\theta/dt$ the rate of cooling of the lower disc, then

$$Q = ms\frac{d\theta}{dt} = (\pi r^2 d'\rho)s\frac{d\theta}{dt},$$

where d' is the thickness of the lower disc and ρ the density of copper. Thus

$$K(\pi r^2)\frac{\theta_1 - \theta_2}{d} = (\pi r^2 d'\rho)s\frac{d\theta}{dt}$$

or $$K = \frac{d'\rho s(d\theta/dt)d}{\theta_1 - \theta_2}$$

Here $d' = 1$ cm, $\rho = 9$ g/cm³,

$s = 0.1$ cal/(g-°C),

$$\frac{d\theta}{dt} = \frac{86-81}{4\times 60} = \frac{1}{48} \ ^\circ\text{C/sec},$$

$$d = 0.48 \text{ cm and } \theta_1 - \theta_2 = 99.5 - 83.5 = 16 \ ^\circ\text{C}$$

$$\therefore \quad K = \frac{1\times 9\times 0.1\times (1/48)\times 0.48}{16}$$

$$= 5.6\times 10^{-4} \text{ cal-cm}^{-1} - \text{sec}^{-1} - {}^\circ\text{C}^{-1}.$$

10. A rubber tube of length 20 cm, through which steam at 100 °C is passing, is immersed in a calorimeter whose water equivalent is 15 gram and which contains 300 gram of water at 16 °C. The temperature of water rises at the rate of 2 °C per minute. If the outer and inner diameters of the tube are 1.0 cm and 0.6 cm respectively, calculate the thermal conductivity of rubber.

Solution : The radial rate of heat-flow through a tube of length l cm and inner and outer radii r_1 and r_2 cm is given by

$$\frac{dQ}{dt} = 2\pi K l \frac{\theta_1 - \theta_2}{\log_e (r_2 - r_1)} \text{cal/sec.}$$

where θ_1 and θ_2 are temperatures inside and outside the tube. If this heat raises the temperature of m gram of water (specific heat 1) contained in a calorimeter of water equivalent w gram by $\Delta\theta$, then

$$\frac{dQ}{dt} = (m+w)\Delta\theta \text{ cal/sec}$$

Equating the above two equations, we get

$$2\pi K l \frac{\theta_1 - \theta_2}{\log_e (r_2/r_1)} = (m+w)\Delta\theta$$

$$\therefore \quad K = \frac{(m+w)\Delta\theta \log_e (r_2 - r_1)}{2\pi l(\theta_1 - \theta_2)}$$

Here $m + w = 300 + 15 = 315$ gram,

$$\Delta\theta = 2\ °C \text{ per minute} = \left(\frac{1}{30}\right)°C/\text{sec},$$

$$\log_e (r_2/r_1) = 2.3 \log_{10} (1.0/0.6) = 2.3 \times 0.2219,$$

$$l = 20 \text{ cm}, \ (\theta_1 - \theta_2) = 100 - 16 = 84\ °C.$$

$$\therefore \quad K = \frac{315 \times (1/30) \times (2.3 \times 0.2219)}{2 \times 3.14 \times 20 \times 84}$$

$$= 5 \times 10^{-4} \text{ cal-cm}^{-1}\text{-sec}^{-1}\text{- }°C^{-1}.$$

11. A heating-wire of 0.0005 meter diameter is embedded along the axis of a cylinder of 0.12 meter diameter. When a current is passed through the wire it gives out a power of 3 kilowatt per meter of its length. If the temperature of the wire be 1500 °C and that of the outer surface of the cylinder be 20 °C, compute the thermal conductivity of the material of the cylinder.

Solution: Let l meter be the length of the wire. Then the power being generated

$$= 3\,l \text{ kilowatt} = 3000\,l \text{ watt}$$

$$= 3000\,l \text{ joule/sec} \qquad (1 \text{ watt} = 1 \text{ joule/sec})$$

$$= \frac{3000}{4.18}\,l \text{ cal/sec}. \qquad (1 \text{ cal} = 4.18 \text{ joule})$$

The rate of radial heat-flow through a cylinder of length l and internal and external radii r_1, and r_2 is given by

$$\frac{dQ}{dt} = 2\pi K l \frac{\theta_1 - \theta_2}{2.3026 \log_{10} (r_2/r_1)}$$

Here $\frac{dQ}{dt} = \frac{3000}{4.18} l \text{ cal/sec}, \theta_1 = 1500°C, \theta_2 = 20\ °C, r_1 = 0.0005$ meter and $r_2 = 0.12$ meter, so that $\frac{r_2}{r_1} = \frac{0.12}{0.0005} = 240.$ Therefore

$$\frac{3000}{4.18}l = 2\times3.14\times K\times l\times\frac{1500-20}{2.3026\log_{10}240}$$

or
$$\frac{3000}{4.18} = 2\times3.14\times K\times\frac{1480}{2.3026\times2.3802}$$

$\therefore$
$$K = \frac{3000\times2.3026\times2.3802}{4.18\times2\times3.14\times1480}$$

$$= 0.423 \text{ cal/(meter = sec-°C)}.$$

12. A copper plate of radius 12 cm and thickness 5 cm conducts when the temperature difference between the circular end faces is 20 °C. A spherical shell of aluminium of radii 9 cm and 4 cm conducts heat radially under the same temperature difference. Compare the amounts of heat transferred per second in the two cases. Thermal conductivities of copper and aluminium are 0.9 and 0.5 C. G. S. units.

Solution: The *linear* rate of the heat-flow through the circular copper plate of area $A(=\pi r^2)$ and thickness d is

$$\left\{\frac{dQ}{dt}\right\}_{plate} = KA\frac{\theta_1-\theta_2}{d}$$

$$= \frac{0.9\times\pi\times(12)^2\times20}{5} \quad \text{...(i)}$$

The *radial* rate of heat-flow through a spherical shell of radii r_1 and r_2 is

$$\left\{\frac{dQ}{dt}\right\}_{shell} = \frac{4\pi K r_1 r_2(\theta_1-\theta_2)}{r_2-r_1}$$

$$= \frac{4\times\pi\times0.5\times4\times9\times20}{5} \quad \text{...(ii)}$$

Comparing the above two expression :

$$\frac{(dQ/dt)_{plate}}{(dQ/dt)_{shell}} = \frac{0.9\times(12)^2}{4\times0.5\times4\times9} = 1.8$$

13. Two thin concentric spherical shells of radii 5 cm and 10 cm respectively have their annular cavity filled with charcoal powder. When energy is supplied at the rate of 10.5 watt to a heater at the centre, a temperature difference of 60 °C is set up between the shells. Find the thermal conductivity of charcoal. (J = 4.2 joule/cal.)

Solution. The thermal conductivity of the charcoal will be given by

$$K = \frac{(dQ/dt)(r_2 - r_1)}{4\pi r_1 r_2 (\theta_1 - \theta_2)}$$

where dQ/dt is the rate of heat supply.

Here radius of one shell $r_1 = 5$ cm,

radius of other shell $r_2 = 10$ cm,

temperature difference $(\theta_1 - \theta_2) = 60$ °C,

and heat supplied per sec, dQ/dt = energy dissipated in heater per sec,

$$= 10.5 \text{ watt}$$

$$= 10.5 \text{ joule/sec}$$

$$= \frac{10.5}{4.2} \text{cal/sec.}$$

Hence from the above expression, we get

$$K = \frac{\frac{10.5}{4.2} \times (10 - 5)}{4 \times 3.14 \times 10 \times 5 \times 60}$$

$$= \frac{10.5 \times 5}{4.2 \times 4 \times 3.14 \times 10 \times 5 \times 60}$$

$$= 3.3 \times 10^{-4} \text{ cal-cm}^{-1} - \text{sec}^{-1} - {}^\circ\text{C}^{-1}.$$

14. In a periodic flow of heat along an iron bar, the periodic time is 4 minutes. If the temperature maximum travels 6 cm in one minute, calculate the thermal conductivity of iron. Density of iron = 7.8 g/cm³, sp. heat of iron = 0.11 cal/(g-°C).

Solution: The velocity of heat-wave in the iron bar is given by

$$v = \sqrt{\frac{4\pi k}{T}},$$

where k is the diffusivity of the bar and T is the period of the wave. From this, we have

$$k = \frac{v^2 T}{4\pi}$$

Further $k = \frac{K}{\rho c}$, where K is conductivity, ρ is density and c is specific heat.

$$\therefore \qquad K = k\rho c = \frac{v^2 T \rho c}{4\pi}$$

Here v = 6 cm/minute = 0.1 cm/sec , T= 4 min = 240 sec, ρ = 7.8 g/cm³ and c = 0.11 cal/(g-°C).

$$\therefore \qquad K = \frac{(0.1\,\text{cm/sec})^2 (240\,\text{sec})(7.8\,\text{g/cm}^3)\left(0.11\frac{\text{cal}}{\text{g-°C}}\right)}{4 \times 3.14}$$

$$= 0.164 \text{ cal/(cm-sec-°C)}$$

6

The Potentials

Thermodynamic Functions

The state of a thermodynamic system of constant mass may be specified in terms of the variables p, V, T, U and S, between which two thermodynamic relations exist:

$$dQ = dU + pdV$$

and $$dQ = TdS.$$

For a complete description of the behaviour of such systems some other relations are required. These relations are mathematically simplified when certain functions of the above variables are introduced. There are four such functions, known as 'thermodynamic functions' described below :

Internal Energy (U) : The internal energy U of a system is a thermodynamic variable. It is also considered as a thermodynamic function.

Suppose a system undergoes an infinitesimal reversible change from one equilibrium state to another. The internal energy changes by an amount

$$dU = dQ - pdV$$

$$= TdS\text{-}pdV.$$

Taking the partial differentials of U, we get

$$\left(\frac{\partial U}{\partial S}\right)_V = T$$

and
$$\left(\frac{\partial U}{\partial V}\right)_S = -p.$$

But dU is a perfect differential, so that

$$\frac{\partial}{\partial V}\left(\frac{\partial U}{\partial S}\right) = \frac{\partial}{\partial S}\left(\frac{\partial U}{\partial V}\right)$$

$$\therefore \qquad \left(\frac{\partial T}{\partial V}\right)_S = -\left(\frac{\partial p}{\partial S}\right)_V \qquad \text{...(i)}$$

This is Maxwell's first thermodynamic relation.

Helmholtz Function (F): It is also called as "Helmholtz Free Energy" or "Thermodynamic Potential at Constant Volume". It is defined by the equation

$$\boxed{F = U - TS}$$

Since U, T and S are perfect differentials, F is also a perfect differential. When a system undergoes an infinitesimal reversible change from an initial equilibrium state to a final equilibrium state, the Helmholtz Free Energy changes by an amount given by

$$dF = dU - TdS - SdT.$$

But $dU = T\,dS - p\,dV$, as shown in the above case.

$$\therefore \qquad dF = (TdS - p\,dV) - TdS - S\,dT$$

$$= -SdT - pdV.$$

Taking the partial differentials of F, we have

$$\left(\frac{\partial F}{\partial T}\right)_V = -S$$

and $$\left(\frac{\partial F}{\partial V}\right)_T = -p$$

Since dF is a perfect differential, we have

$$\frac{\partial}{\partial V}\left(\frac{\partial f}{\partial T}\right) = \frac{\partial}{\partial T}\left(\frac{\partial f}{\partial V}\right)$$

$$\left(\frac{\partial S}{\partial V}\right)_T = \left(\frac{\partial p}{\partial T}\right)_V$$

This is Maxwell's second relation.

Enthalpy or Total Heat Function (H): This is defined by

$$\boxed{H = U + pV}$$

For an infinitesimal reversible change, we get

$$dH = dU + pdV + Vdp.$$

But $dU + pdV = dU + dW = dQ$ (by first law) $= TdS$ (by second law)

$$\therefore \qquad dH = TdS + Vdp$$

The partial differentials of H are

$$\left(\frac{\partial H}{\partial S}\right)_p = T \quad \text{and} \quad \left(\frac{\partial H}{\partial p}\right)_S = V$$

Since dH is a perfect differential, we have

$$\frac{\partial}{\partial p}\left(\frac{\partial H}{\partial S}\right) = \frac{\partial}{\partial S}\left(\frac{\partial H}{\partial p}\right)$$

$$\therefore \qquad \left(\frac{\partial T}{\partial p}\right)_S = \left(\frac{\partial V}{\partial S}\right)_p \qquad \text{...(iii)}$$

This is Maxwell's third relation.

Gibbs Function (G) : This is also known as "Gibbs Free Energy " or "Thermodynamic Potential at Constant Pressure". It is defined by

$$\boxed{G = H - TS}$$

For an infinitesimal reversible process.

$$dG = dH - TdS - SdT$$

But $dH = TdS + Vdp$ (as shown above).

$$\therefore \qquad dG = Vdp - SdT.$$

The partial differentials of G are

$$\left(\frac{\partial G}{\partial p}\right)_T = V$$

and
$$\left(\frac{\partial G}{\partial T}\right)_p = -S$$

Since dG is a perfect differential, we have

$$\frac{\partial}{\partial T}\left(\frac{\partial G}{\partial P}\right) = \frac{\partial}{\partial p}\left(\frac{\partial G}{\partial T}\right)$$

$$\therefore \qquad \left(\frac{\partial V}{\partial T}\right)_P = -\left(\frac{\partial S}{\partial p}\right)_T \qquad \ldots(iv)$$

This is Maxwell's fourth relation.

Importance : Thermodynamic functions are of practical importance in studying the conditions of equilibrium of a system. For example, the condition of equilibrium for a process in which the temperature and volume of the system remain constant (isothermal-isochoric process) may be expressed as

$$dF = 0. \qquad [\because dF = -pdV - SdT]$$

This means that out of the various states which a system can assume by isothermal-isochoric process, only those are stable in which the Helmholtz Free Energy F (= $U - T\,S$) is a minimum. The equation $dF = 0$ refers to a minimum value of F (and not to a maximum value). It follows from the fact that in any natural (irreversible) process the Helmholtz Free Energy can only fall because the entropy can only increase.

Similarly, the condition of equilibrium for a process in which the temperature and pressure of the system remain constant (isothermal-isobaric process) may be expressed as

$$dG = 0. \qquad [\because dG = Vdp - SdT]$$

This means that a system at constant temperature and pressure is in stable equilibrium when the Gibbs function G (the thermodynamic potential at constant pressure) is a minimum.

Enthalpy : The enthalpy H of a thermodynamic system is defined as

$$H = U + pV$$

When the system undergoes an infinitesimal process from an initial equilibrium state to a final equilibrium state, the change in enthalpy is

$$dH = dU + pdV + Vdp.$$

But $dU + pdV = dQ$ (from first law).

$$\therefore \qquad dH = dQ + Vdp.$$

For an isobaric (pressure remaining constant) process, $dp - 0$, so that

$$dH = dQ.$$

If H_i and H_f represent the initial and final enthalpies of the system, then we have

$$H_f - H_i = Q.$$

Thus, *the change in enthalpy during an isobaric process is equal to the heat transferred* . This result is useful in engineering and chemistry where so many processes take place at atmospheric (that is, at constant) pressure.

Enthalpy in Porous Plug Experiment : An important property of the enthalpy function is in connection with a throttling process (porous plug experiment). In this process a gas is made to pass under a constant pressure through an insulated porous plug to a region of lower constant pressure. Let p_1 and V_1, be the initial pressure and volume and p_f and V_f the final pressure and volume of the gas.

The *net* external work done by the gas is $p_f V_f - p_i V_i$, Since there is no heat-exchange between the gas and the surroundings, this work is done at the cost of internal energy of the gas. Thus if U_i and U_f be the initial and final internal energies, we have

$$U_i - U_f = p_f V_f - p_i V_i$$

or
$$U_i + p_i V_i = U_f + p_f V_f$$

But $U + pV$ is defined as enthalpy H of the system.

$$\therefore \qquad H_i = H_f$$

Thus, in a throttling process the initial and final enthalpies are equal.

Isoenthalpic Curve

An isoenthalpic curve is the locus of all points representing equilibrium states of the same enthalpy of a thermodynamic system.

An isoenthalpic curve for a gas can be drawn by means of porous-plug experiment. The (initial) pressure and temperature p_i, and T_i on the high-pressure side of the plug are arbitrarily selected. The (final) pressure p_f on the other side of the plug is set at any value smaller than P_i and the temperature T_f of the gas is measured. This is done for a number of different values of p_f. Thus, we get a number of points 1,2, 3, ... for the various (p_f, T_f) values corresponding to a point (p_i, T_f). All these points represent equilibrium states of some constant mass of the gas, at which the gas has the same enthalpy. The locus of all these discrete points is the isoenthalpic curve.

Other curves corresponding to different enthalpies can be obtained by selecting different values of $T_{i'}$.

The numerical value of the slope of an isoenthalpic curve at any point is the Joule-Kelvin coefficient at that point. The locus of all points at which the Joule-Kelvin coefficient is zero (*i.e.* the locus of the maxima of the curves) is the "inversion curve".

Isothermal Curve for an Ideal Gas is Isoenthalpic : By definition, the enthalpy of a thermodynamic system is

$$H = U + pV.$$

For an ideal gas $pV = RT$, so that

$$H = U + RT.$$

For an isothermal (T constant) process, the internal energy U of an ideal gas is constant, because it depends *only* on the (constant) temperature. Hence , enthalpy H is also constant. It means that a p-V curve for an ideal gas at constant temperature will be a locus of constant enthalpy, that is, it will be isoenthalpic.

Internal Energy and Free Energy : *The internal energy U of a system is the energy which it possesses due to its molecular constitution and motion.* In general, it is the sum of the kinetic energy of the molecules due to their motion and the potential energy of the molecules due to their mutual attraction.

The free energy of a system has been defined by

$$F = U - TS, \qquad ...(i)$$

where T is the Kelvin temperature and S the entropy of the system. Suppose the system undergoes an infinitesimal reversible change. Then, the change in F will be given by

$$dF = dU - (TdS + SdT).$$

But $dU = dQ - dW$ (I law) and $dQ = TdS$ (II law); so that $dU = TdS - dW$.

$$\therefore \quad dF = (TdS - dW) - (TdS + SdT)$$

$$= - dW - SdT. \qquad ...(ii)$$

If the change be isothermal ($dT = 0$), then

$$dF = - dW. \qquad ...(iii)$$

Thus , the work done is exactly equal to the change in the free energy. This means that in a reversible isothermal change, the external work dW is done wholly at the cost of the free energy of the system. Hence , *the free energy of a system is the energy which is available for work in reversible isothermal change.* Thus, it corresponds exactly to the potential energy for mechanical system.

Rewriting eq. (i), we get

$$U = F+TS. \qquad ...(iv)$$

This show that the internal energy U of a system consists of two parts : (i) the free energy F which is available for work in reversible isothermal changes ; and (ii) the latent (or bound) energy $T\ S$ which is unavailable for useful work. As entropy S increases, the unavailable energy also increases and therefore the available energy decreases.

Gibbs-Helmholtz Equation

From eq. (ii), we have

$$dF = - dW - SdT$$

$$= -pdV - SdT.$$

Writing the partial differential of F at constant volume $(dV = 0)$; we have

$$\left(\frac{\partial F}{\partial T}\right)_V = -S$$

Substituting this value of S in eq. (iv), we get

$$\boxed{U = F - T\left(\frac{\partial F}{\partial T}\right)_V}$$

This is Gibbs-Helmholtz equation.

Importance : This equation does not involve the entropy S whose calculation is often difficult. Hence it can easily be applied to study the thermodynamics of isothermal changes in a chemical system.

Suppose a chemical system changes isothermally at temperature T from state 1 to state 2. Then , from Gibbs-Helmholtz equation, we have

$$U_2 - U_1 = F_2 - F_1 - T\frac{\partial}{\partial T}(F_2 - F_1)$$

or $$\Delta U = \Delta F - T\left[\frac{\partial(\Delta F)}{\partial T}\right]_V$$

But from eq. (iii), $-\Delta F = W$, which is the work obtainable from the system during a reversible change, and from the first law of thermodynamics $-\Delta U = U_r$, which is the heat of reaction at constant volume.

$\therefore$ $$-U_r = -W - T\left[\frac{\partial}{\partial T}(-W)\right]$$

or $$W - U_r = T\left(\frac{dW}{dT}\right).$$

Thus we can calculate, the variation of W with temperature, *i.e.* we can calculate the temperature at which the required amount of work would be obtained from the system.

Gibbs Function (or Gibbs Free Energy): The Gibbs function G of a thermodynamic system is defined as

$$G = H - TS, \qquad \text{...(i)}$$

where H is the enthalpy and S is the entropy of the system.

For an infinitesimal reversible process , the change in the Gibbs function is

$$dG = dH - TdS - SdT.$$

But $dH = T\,dS + V\,dp$

$$\therefore \qquad dG = Vdp - SdT. \qquad \text{...(ii)}$$

If during the reversible process, the temperature remains constant $(dT = 0)$ and also the pressure remains constant $(dp = 0)$, then , we have

$$dG = 0$$

or

$$G = \text{constant}.$$

Thus, for changes at constant temperature and constant pressure (isothermal-isobaric process like fusion, sublimation, vaporisation etc.) the Gibbs function remains constant.

If the process is irreversible (natural) the Gibbs function G falls (because entropy S increases). Hence we may conclude that for changes at constant temperature and constant pressure, the Gibbs function falls or (in the limiting case) remains constant. It never increases.

Now, writing the partial differential of G at constant pressure $(dp = 0)$, from eq. (ii), we have

$$\left(\frac{\partial G}{\partial T}\right)_p = -S$$

Making this substitution in eq. (i), we get

$$G = H + T\left(\frac{\partial G}{\partial T}\right)_P$$

PROBLEM

1. One gram of water when converted to steam at atmospheric pressure occupies a volume of 1671 cm^3. The latent heat of vaporisation at this temperature is 539 cal/g. Compute the increase in internal energy (ΔU), entropy (ΔS), enthalpy (ΔH) and the Gibbs function (ΔG) when one gram of water is evaporated at this temperature and pressure. Express all answers in the same system of units.

Solution : The heat taken in by 1 gm of water in converting to steam at 100 °C is given by

$$\Delta Q = mL = 1 \times 539 = 539 \text{ cal}$$

$$= 539 \times (4.18 \times 10^7)$$

$$= 22.53 \times 10^9 \text{ erg.}$$

The conversion takes place at *constant pressure*, and so the work done is

$$\Delta W = pdV$$

$$= (1.01 \times 10^6 \text{ dynes/cm}^2)(1671-1) \text{ cm}^3$$

$$= 1.69 \times 10^9 \text{ erg.}$$

Therefore, the change in internal energy is

$$\Delta U = \Delta Q - \Delta W$$

$$= 22.53 \times 10^9 - 1.69 \times 10^9$$

$$= 20.84 \times 10^9 \text{ erg.}$$

The change in entropy is

$$\Delta S = \frac{\Delta Q}{T} = \frac{22.53 \times 10^9}{373} = 6.04 \times 10^7 \text{ erg/K}$$

For the change in enthalpy, we write

$$H = U + pV$$

$$\Delta H = \Delta U + p\,\Delta V + V\,\Delta p$$

$$= \Delta U + p\,\Delta V$$

$$[\Delta p = 0 \text{ , at constant pressure}]$$

$$= \Delta U + \Delta W$$

$$= \Delta Q$$

$$= 22.53 \times 10^9 \text{ erg.}$$

The change in Gibbs function in an isothermal-isobaric process is zero

$$\Delta G = 0.$$

7

Event Probability

Let us consider the experiment of tossing a coin. There are only two possible outcomes; either the 'head' or the tail' falls uppermost. Suppose, we toss the coin a very *large* number of times and count the fraction of cases when the outcome is a head and that when it is a tail. These fractions then give respectively the measured probability p of obtaining a head and the measured probability q of obtaining a tail. Common experience tells us that the events showing the head uppermost and the events showing the tail uppermost are half and half. That is, the probability of each event is $\frac{1}{2}$ $(p = q = \frac{1}{2})$. Thus, *the probability of occurrence of an event is the ratio of the number of cases in which the event occurs to the total number of cases, provided the total number of cases is quite large.*

If an event occurs in a ways and fail to occur in b ways, and each of these ways is equally likely, then the probability of its occurring is $\frac{a}{a+b}$ and the probability of its failing is $\frac{b}{a+b}$.

The sum of the two probabilities $\left(\frac{a}{a+b}+\frac{b}{a+b}\right)$. is, however, 1 because the event must either occur or fail.

Let us now consider the case of two *distinguishable* coins marked 1 and 2 (say, a silver coin and a copper coin). Suppose we toss the coins a number of times and count the following events :

(i) Heads of both fall uppermost ...a_1a_2

(ii) Tails of both fall uppermost ...b_1b_2

(iii) Head of 1 and tail of 2 uppermost ...a_1b_2

(iv) Tail of 1 and head of 2 uppermost ...b_1a_2

Here a denotes head, b denotes tail, a_1 signifies that head of 1 is uppermost, and so on.

The possible events are the terms of the product $(a_1 + b_1)(a_2 + b_2) = a_1 a_2 + a_1 b_2 + b_1 a_2 + b_1 b_2$, and the total number of events is equal to the number of terms (4) of the product.

The events are independent and the probability of each event is 1/4 ; because all the four events are equally likely. This can be considered as follows :

If we toss 1, the probability of getting the head a_1 is 1/2; if we toss 2, the probability of getting the head a_2 is also 1/2. Clearly, in a simultaneous toss of both the coins, the probability of getting the heads of both, a_1a_2, is $(1/2) \times (1/2) = 1/4$; because there are in all 4 equally-likely ways in which the composite event takes place and the desired event takes place only once. Thus, *the probability of a composite event (here $a_1 a_2$) is the product of probability of the individual, independent events (here a_1 and a_2).*

Let us now consider the case of two coins which are *indistinguishable* from each other (say, both are silver coins). Now the events (iii) and (iv) above can no more be distinguished from each other and as such there are only three distinguishable events:

(i) Heads of both coins uppermost .. .a^2

(ii) Tails of both coins uppermost ...b^2

(iii) Head of one and tail of other uppermost ...ab.

One can obtain these events in the terms of the binomial expansion of $(a + b)^2$, *i.e.*,

$$(a + b)^2 = a^2 + 2ab + b^2.$$

The probability of getting each of the events a and b is 1/4 , while that of the event ab is 2 x (1/4) = 1/2. Thus, *the probability of any event is proportional to the coefficient of the term representing that event in the binomial expansion, or the relative probabilities are in the ratio of the coefficients of the corresponding terms in the binomial expansion.*

Extension to N Coins

Let us now consider the general case of N identical coins tossed simultaneously a number of times. The possible events are given by the terms of the binomial expansion of

$$(a+b)^N,$$

that is,

$$a^N,\ {}^NC_1\, a^{N-1}\, b,\ {}^NC_2\, a^{N-2}\, b^2, \ldots,\ {}^NC_r\ a^{N-r}\ b^r\ , \ldots, b^N.$$

The term $a^{N-r}\, b^r$ means that $(N - r)$ coins have heads uppermost and r coins have tails uppermost, and this event can occur in ${}^NC_r$ ways. The number of ways for the occurrence of the converse event $a^r\, b^{N-r}$ (r heads, N-r tails) is also ${}^NC_r$ (because ${}^NC_r = {}^NC_{N-r}$). In other words, the relative probability, or possibility number, of the event $a^r\, b^{N-r}$ is

$${}^NC_r = \frac{N!}{r!(N-r)!}$$

The total number of ways of all the events is

$$\sum_{r=0}^{N} {}^N C_r = (1+1)^N = 2^N.$$

Therefore, the (mathematical) probability of the event $a^r b^{N-r}$ is

$$P(r) = \frac{{}^N C_r}{\sum_{r=0}^{N} {}^N C_r} = \frac{N!}{r!(N-r)!} \frac{1}{2^N}$$

For a given N, the probability is a function of r and is called the 'binomial distribution'.

In the case of a single toss of a coin there are only two possible events (either the head up or the tail up) and the probability of each is ½. If, however, the probabilities of the two events are different, say p and q, then, the above formula will be generalised to the form

$$P(r) = \frac{N!}{r!(N-r)!} p^r q^{N-r}.$$

This reduces to $\dfrac{N!}{r!(N-r)!} \dfrac{1}{2^N}$, if $p = q = 1/2$.

Extension to the case of N Dice: A coin has only two states, head or tail. We now proceed to bodies having a large number of possible states. An example is the game of dice. Each die has six faces which we may denote by a, b, c, d, e, f. When we toss N dice, we get a combination

$$(a)^{n_1} (b)^{n_2} (C)^{n_3} (d)^{n_4} (e)^{n_5} (f)^{n_6}$$

that is, n_1 dice fall with face 'a' uppermost, n_2 dice fall with face 'b' uppermost, and so on, and $N = n_1 + n_2 + n_3 + n_4 + n_5 + n_6$. Proceeding in the same manner as for the case of coins, the probability of occurrence of the above combination is

$$P = \frac{N!}{n_1!\, n_2!\, n_3!\, n_4!\, n_5!\, n_6!} \frac{1}{6^N}.$$

Probability : Definition : *The probability of occurrence of an event is the ratio of the number of cases in which the event occurs to the total number of cases, provided the total number of cases is quite large.*

For example, suppose a coin is tossed a large number n times. Then, the probability of getting head is

$$P(h) = \frac{\text{number of times head is obtained}}{\text{total number of times the coin is tossed}} = \frac{n_h}{n}.$$

This ratio tends to be $\frac{1}{2}$ as one tosses the coin a large number of times $(n \to \infty)$.

Probability Relations : There are two important probability relations or probability theorems:

Addition Theorem : *For mutually exclusive events, if the probability of occurrence of one event r is P(r) and that of the other event s is P(s), then the probability of occurrence of either r or s is P(r)+ P(s).*

Suppose, we have a large number N of similar systems, and experiment on any system can lead to any of the mutually exclusive results (or events) 1, 2, 3,........n. Let N_1 systems lead to result 1, N_2 to result 2,........ and N_n to result n. Since these n results are mutually exclusive, it follow that

$$N_1 + N_2 + \ldots\ldots + N_i + N_n = N.$$

Dividing by N, we get

$$\frac{N_1}{N} + \frac{N_2}{N} + \ldots\ldots + \frac{Ni}{N} + \ldots\ldots + \frac{N_n}{N} = 1$$

or $$P(1) + P(2) + \ldots\ldots + P(i) + \ldots\ldots + P(n) = 1. \quad \ldots(i)$$

Here $P(i) \left(\equiv \frac{N_i}{N} \right)$ represents the probability of occurrence of result i. Eq. (i) means that the sum of all the possible probabilities equals unity. This is called 'normalization condition' for probabilities, and can be written as

$$\sum_{i=1}^{n} P(i) = 1$$

Out of N systems, suppose N_r systems exhibit result r and N_s systems exhibit result s. Thus, there are $(N_r + N_s)$ systems exhibiting either result r or result s. Correspondingly, the probability $P\ (r\ or\ s)$ of occurrence of either result r or s is given by

$$P(r \text{ or } s) = \frac{N_r + N_s}{N} = \frac{N_r}{N} + \frac{N_s}{N}.$$

or

$$\boxed{P(r \text{ or } s) = P(r) + P(s).}$$

This result may be generalised to any number of results (events).

Thus, for n mutually exclusive events, we have

$$\text{P(1 or 2 or ...... } n) = P(1) + P(2) + ... + P(n).$$

Multiplication Theorem : If the probabilities of occurrence of two independent events are P(r) and P(s), then the probability of occurrence of both of them simultaneously is P(r) P(s).

Suppose that each of the N similar systems under consideration can exhibit two different types of results labelled by r (where r = 1, 2, 3,...) and by s (where s = 1, 2, 3,...). Let $P(rs)$ denote thy probability of joint occurrence of *both* event r *and* event s simultaneously. Let, out of N systems, N_{rs} exhibit jointly *both* an event r of the first type and an event s of the second type. Then

$$P(rs) \equiv \frac{N_{rs}}{N}.$$

Suppose that the probability of occurrence of the result r is unaffected by the occurrence or non-occurrence of the result s. The results of type r and s are then said to be statistically independent. Now, suppose N_r systems exhibit any particular result r, and a fraction *P(s)* of these systems also exhibit the result *s*. Then, the number of systems exhibiting jointly both r and *s* is :

$$N_{rs} = N_r \, P(s).$$

Correspondingly, the probability of joint occurrence of *both r* and *s is*

$$P(rs) \equiv \frac{N_{rs}}{N} = \frac{N_r \, P(s)}{N} = P(r) \, P(s).$$

Hence we conclude that if the events *r* and *s* are 'statistically independent', then

$$\boxed{P(rs) = P(r) \, P(s).}$$

Tossing of N Coins : Let us consider *N* identical coins which are tossed simultaneously a number of times. We wish to have r heads uppermost, which also means *(N - r)* tails uppermost.

The number of ways in which *r* heads (and *N-r* tails) can occur in N identical coins is

$$^N C_r = \frac{N!}{r! \, (N-r)!}$$

The total number of ways of all possible combinations is

$$\sum_{r=0}^{N} {}^N C_r = (1+1)^N = 2^N .$$

∴ probability of having r heads up (and *N-r* tails) is

$$P(r) = \frac{^N C_r}{\sum_{r=0}^{N} {}^N C_r} = \frac{N!}{r! \, (N-r)!} \frac{1}{2^N}. \quad \text{...(i)}$$

Most Probable Combination : The combination of heads and tails which is most likely to occur is the one for which *P(r)* is maximum. Now, *P(r)* is maximum when C_r is maximum.

From Algebra, ${}^N C_r$ is maximum when $r = N/2$. Thus, *the most probable combination is the one having as many heads as the tails.*

Putting $r = N/2$ in eq. (i), we get

$$P_{max} = \frac{N!}{\frac{N}{2}!\frac{N}{2}!}\frac{1}{2^N}.$$

Least Probable Combination : The combination of heads and tails least likely to occur is the one having minimum $P(r)$. Again, the probability $P(r)$ is minimum when ${}^N C_r$ is minimum, that, is when r=0 or N. Thus, *the least probable combination is the one having all heads or all tails.*

Putting $r = 0$ or $r = N$ in eq. (i). we get

$$P_{min} = \frac{N!}{0!\,N!}\frac{1}{2^N}.$$

$$= \frac{1}{2^N}. \qquad [\because 0! = 1]$$

System of N Particles

N particles distribute themselves between two states A and B. If r particles are in the state A, then there will *be* N-r particles in the state B.

The number of ways in which r particles can be chosen from N particles is

$${}^N C_r = \frac{N!}{r!\,(N-r)!}.$$

Now, the probabilities for a particle to be found in the states A and B are p and (1- p) respectively. Then, by multiplication theorem, the probability for r particles to be found in the state A will be $(p)^r$ and that for $(N$ - r) particles to be found in the state B will be $(1\text{-}p)^{N\text{-}r}$.

$\therefore$ probability in which r particles are in state A and $(N - r)$ in state B is

$$P(r, N-r) = {}^{N}C_r(p)^r (i - p)^{n-r}$$

$$= \frac{N!}{r!\,(N-r)!}(p)^r (1-p)^{N-r}.$$

Deviation from Most Probable Distribution: Let us consider a system of $2N$ particles which distribute themselves in two identical boxes. If one box contains r particles, the other will have $2N - r$ particles.

The number of ways in which r particles can be chosen from $2N$ particles is

$${}^{2N}C_r = \frac{2N!}{r!\,(2N-r)!}.$$

The total number of possible distributions from $(2N, 0)$ to $(0, 2N)$ is

$$\sum_{r=0}^{2N} {}^{2N}C_r = 2^{2N}$$

Therefore, the probability for the distribution in which r particles go to one box and $(2N - r)$ to the other is

$$P(r, 2N-r) = \frac{{}^{2N}C_r}{\sum {}^{2N}C_r} = \frac{2N!}{r\,!\,(2N-r)!}\,\frac{1}{2^{2N}}. \qquad \text{...(i)}$$

The most probable distribution corresponds to equal number of particles in the two boxes i.e. $r = N$.

$$P(N, N) = \frac{2N!}{N\,!\,N!}\,\frac{1}{2^{2N}}. \qquad \text{...(ii)}$$

Let us now consider a distribution *slightly* deviated from the most probable distribution, having $N + r$ particles in one box and N-r particles in the other where $r/N << 1$. According to eq. (i), the probability of this distribution will be

$$P(N+r, N-r) = \frac{2N!}{(N+r)!\,(N-r)!}\,\frac{1}{2^{2N}}. \qquad \text{...(iii)}$$

The ratio of this distribution to the most probable distribution is

$$R = \frac{P(N+r, N-r)}{P(N,N)}$$

$$= \frac{N!\,N!}{(N+r)!\,(N-r)!}$$

$$= \frac{N!}{(N+r)!}\,\frac{N!}{(N-r)!}$$

$$= \frac{N!}{(N+r)(N+r-1)(N+r-2)...(N+1)[N!]}\,\frac{N(N-1)(N-2)...(N-r+1)[(N-r)!]}{(N-r)!}$$

$$= \frac{N(N-1)(N-2)...(N-r+1)}{(N+r)(N+r-1)(N+r-2)...(N+1)}$$

$$= \frac{N}{N+r}\,\frac{N-1}{N+r-1}\,.........\,\frac{N-r+2}{N+2}\,\frac{N-r+1}{N+1}$$

$$= \left\{\frac{N}{N+r}\right\}\left\{\frac{N-1}{(N-1)+r}\right\}.........\left\{\frac{(N+2)-r}{N+2}\right\}\left\{\frac{(N+1)-r}{N+1}\right\}$$

$$= \left\{1-\frac{r}{N+r}\right\}\left\{1-\frac{r}{(N-1)+r}\right\}.........\left\{1-\frac{r}{N+2}\right\}\left\{1-\frac{r}{N+1}\right\}$$

For $r/N << 1$ or $r << N$ and $N >> 1$, this can be written as

$$R = \left(-\frac{r}{N}\right)\left(1-\frac{r}{N}\right)..........\left(1-\frac{r}{N}\right)\left(1-\frac{r}{N}\right)$$

$$= \left(1-\frac{r}{N}\right)^r.$$

Taking natural logarithm of both sides :

$$\log_e R = r \log_e \left(1 - \frac{r}{n}\right)^*$$

$$= r\left(-\frac{r}{N}\right) = -\frac{r^2}{N} \qquad \left(\text{neglecting } \frac{r^2}{N^2} \text{ etc.}\right)$$

Putting $r/N = f$ (fractional deviation from the most probable distribution), we have

$$\log_e R = -f^2 N$$

or $$\boxed{R = e^{-f^2 N},} \qquad \text{where } f = r/N$$

This is the relative probability of distribution corresponding to a fractional deviation f from the most probable distribution. This shows that *the probability of distribution narrows down with increasing N.*

Graphs of relative probability R against the fractional deviation f for various values of N.

As an example, let $N = 10^7$. For a distribution which deviates by only 1 part in 10^3 (*i.e.* $f = 10^{-3}$), we have

$$f^2 N = (10^{-3})^2\, 10^7 = 10$$

$$\therefore \quad R = e^{-10} \simeq 4 \times 10^{-5}$$

The probability drops down to about 10^{-5} of the most probable distribution.

Probability of a Distribution

Let us consider a large box divided into k cells of areas a_1, a_2, a_3,.... a_k. We throw N identical (but distinguishable) balls into the box in a completely random manner, so that no part of the box is favoured. We note down the number of balls falling in each cell, and then repeat the experiment a *large* number of times. We find that a particular distribution of balls among the cells occurs more often than any other. We call this the 'most probable distribution'. Obviously, the most probable distribution is one in which the number of balls in each cell is proportional to the size of the cell (a large cell is more likely to be hit than a small cell). Let us obtain this result by actually calculating the probabilities of the various possible distributions.

The probability P that the balls be distributed in a certain way among the cells depends upon two factors :

(i) The *a priori probability* G of the distribution, which is based upon the properties of each cell, and

(ii) the *thermodynamic probability* W of the distribution, which is the number of different sequences in which the balls may be distributed among the cells without changing the number in each cell.

The a priori probability g_i that a ball falls into the ith cell is the ratio of the area a_i of the cell and the total area A of the entire box. That is,

$$g_i = \frac{a_i}{A},$$

where $$A = a_1 + a_2 + \dots\dots\dots\dots + a_k .$$

The sum of the a priori probabilities of all the cells must be 1, since the ball falls somewhere into the box. That is,

$$\sum_i g_i = g_1 + g_2 + \dots\dots\dots + g_k = 1.$$

By multiplication theorem, the probability that two balls fall in the i th cell is $g_i \times g_i = g_i^2$. Therefore, the a priori probability that n_i balls fall in the i th cell is $(g_i)^{n_i}$. Thus, the a priori probability G of any particular distribution of N balls among the k cells such that n_1 balls fall in the first cell, n_2 balls fall in the second cell is

$$G = (g_1)^{n_1} (g_2)^{n_2} \ldots\ldots\ldots (g_i)^{n_i} \ldots\ldots\ldots (g_k)^{n_k} \qquad \ldots(i)$$

subject to the condition

$$\sum_i n_i = n_1 + n_2 + \ldots\ldots + n_k = N.$$

If the cells are of equal size, they all have the same a priori probability g, and $G = g^N$.

Now, all the distributions of balls among the cells are not equally probable, we have to introduce the concept of thermodynamic probability. We have n_1 balls in the first cell, n_2 balls in the second cell,...... and so on. There can be a number of ways in which this configuration can be obtained, and this number is the thermodynamic probability W for this configuration.

The number of ways in which any n_1 balls out of total N balls can fall in the first cell is

$$^{N}C_{n_1} = \frac{N!}{n_1\,!\,(N-n_1)!}.$$

The number of ways in which any n_2 balls out of remaining $(N - n_1)$ balls can fall in the second cell is

$$^{N-n_1}C_{n_2} = \frac{N!}{n_2\,!\,(N-n_1-n_2)!}.$$

......and so on; n_k balls out of remaining $(N - n_1 - n_2 - \ldots - n_{k-1})$ balls can fall in the k th cell in a number of ways.

The total number of ways, W, in which n_1 balls can fall in cell 1, n_2 in cell 2,

......and n_k balls in cell k, is

$$W = \frac{N!}{n_1\,!\,(N-n_1)!} \times \frac{(N-n_1)\,!}{n_2\,!\,(N-n_1-n_2)!} \times \frac{(N-n_1-n_2)!}{n_3\,!\,(N-n_1-n_2-n_3)!}$$

$$\times \ldots\ldots \times \frac{(N-n_1-n_2-\ldots n_{k-1})!}{n_k\,!\,(N-n_1-n_2-n_2\ldots n_k)!}$$

$$= \frac{N!}{n_1!\ n_2!\ n_3!\ .....n_k!} \qquad ...(ii)$$

because $n_1 + n_2 + ... + n_k = N$ and $0! = 1$.

The total probability P of the distribution is the product of the a priori probability G and the thermodynamic probability W given by eq. (i) and (ii). Then

$$\boxed{P = \frac{N!}{n_1!\ n_2!\ .....n_k!}(g_1)^{n_1}(g_2)^{n_2}........(g_k)^{n_k},} \qquad ...(iii)$$

where $g_1 = a_1/A$,.... This is the probability of the distribution; n_1 balls in cell of area a_1, n_2 balls in cell of area a_2, and so on.

The Most Probable Distribution : Let us now find which distribution of the balls has the largest probability P. For mathematical convenience, we consider $\log_e P$ rather than P itself. Taking natural logarithm of eq. (iii), we have

$$\log_e P = \log_e N! - \sum_i^k \log_e n_i! + \sum_i^k n_i \log_e g_i.$$

By Stirling's approximation $\log_e n! = n \log_e n - n$, the last expression may be written as

$$\log_e P = N \log_e N - N - \sum_i n_i \log_e n_i + \sum_i n_i + \sum_i n_i \log_e g_i.$$

But $\sum_i n_i = N$. Therefore

$$\log_e P = N \log_e N - \sum_i n_i \log_e n_i + \sum_i n_i \log_e g_i. \qquad ...(iv)$$

For most probable distribution, small changes δn_i in any of the n_i's do not affect the value of P or $\log_e P$, *i.e.*

$$\frac{\delta(\log_e P_{\max})}{\delta n_i} = 0$$

or

$$\delta(\log_e P_{\max}) = 0$$

Using eq. (iv), this yields

$$\delta(\log_e P_{\max}) = -\sum_i n_i \delta(\log_e n_i) - \sum_i \delta n_i (\log_e n_i) + \sum_i \delta n_i (\log_e g_i) = 0 \quad \text{...(v)}$$

because $\delta(N \log_e N) = 0$, as N is constant. Now

$$\delta(\log_e n_i) = \frac{1}{n_i} \delta n_i$$

and so $$\sum_i n_i \, \delta(\log_e n_i) = \sum_i \delta n_i.$$

But $$\sum_i \delta n_i. = 0 \quad \text{...(vi)}$$

because the total number of balls $N = \sum_i n_i$ is constant.

$$\therefore \quad \sum_i n_i \, \delta(\log_e n_i) = 0$$

Making this substitution in eq. (v), we get

$$-\sum_i \delta n_i (\log_e n_i) + \sum_i \delta n_i (\log_e g_i) = 0. \quad \text{...(vii)}$$

Let us now use the method of Lagrange's undetermined multipliers. We multiply eq. (vi) by a quantity α which does not depend upon any of the n_i's, and get

$$\sum_i \alpha \, \delta n_i = 0$$

Adding it to eq. (vii), we get

$$-\sum_i \delta n_i (\log_e n_i) + \sum_i \delta n_i (\log_e g_i) + \sum_i \alpha \; \delta n_i = 0$$

or $$\sum_i (-\log_e n_i + \log_e g_i + .\alpha) \delta n_i = 0$$

As δn_i's are independent variables, therefore, for the validity of the last equation, the quantity in parentheses must always be zero regardless of the values given to the δn_i's. Thus, we have

$$-\log_e n_i + \log_e g_i + \alpha = 0$$

or $$n_i = g_i\, e^{\alpha}\,. \qquad \text{...(viii)}$$

Since α does not depend upon i, we have

$$\sum_i n_i = e^{\alpha} \sum_i g_i\,.$$

But $\sum_i g_i$ is the sum of the a priori probabilities of all the cells and must be equal to 1, as considered above.

$$\therefore \quad \sum_i n_i = e^{\alpha}$$

But $\sum_i n_i$ is the total number of balls, N.

$$\therefore \qquad e^{\alpha} = N$$

Eq. (viii) therefore becomes

$$n_i = N g_i$$

Thus, for most probable distribution, the number of balls in any cell is proportional both to the total number of balls N and to the a priori probability g_i which is equal to the relative size of the cell.

Since $g_i = \dfrac{a_i}{A}$, therefore

$$n_i = \left(\frac{N}{A}\right) a_i\,.$$

Thus, for most probable distribution, the number of balls in any cell is equal to the product of the average density of balls N/A and the area of the cell.

Maxwell-Boltzmann Distribution Law **:** This law describes how a fixed total amount of energy is distributed among the various members of an assembly of identical but distinguishable particles (such as the molecules of a gas) in the most probable distribution.

Suppose we have an assembly of N molecules whose energies are limited to the k values ε_1, ε_2,....... ε_i,........ ε_k arranged in the order of increasing energy. These energies may represent either discrete energy states or average energies within a sequence of energy intervals. If there are n_1 molecules of energy ε_1, n_2 molecules of energy ε_2, and so on, then the most probable distribution of molecules among these k energies is subject to two conditions :

(1) The total number of molecules N is constant *i.e.*

$$N = n_1 + n_2 + ... + n_i + ...n_k = \text{constant}$$

so that $$\delta N = \delta n_1 + \delta n_2 + ... + \delta n_i + ...\delta n_k = 0$$

or $$\sum_i \delta n_i = 0 \qquad ...(i)$$

(2) The total energy E of the assembly is constant *i.e.*

$$E = \varepsilon_1 n_1 + \varepsilon_2 n_2 + ... + \varepsilon_i n_i + ...\varepsilon_k n_k = \text{constant}$$

or $$\delta E = \varepsilon_1 \delta n_1 + \varepsilon_2 \delta n_2 + ... + \varepsilon_i \delta n_i + ...\varepsilon_k \delta n_k = 0$$

or $$\sum_i \varepsilon_1 \delta n_i = 0 \qquad ...(ii)$$

If the a priori probability for a molecule to have the energy ε_1 is g_i the probability P for any distribution is

$$P = \frac{N!}{n_1!\, n_2!\, ...n_k!} (g_1)^{n_1} (g_2)^{n_2} .....(g_k)^{n_k}.$$

Taking natural logarithm of this equation, and using Stirling's approximation, we can show (as before) that the most probable distribution must obey the following equation :

$$-\sum_i \delta n_i (\log_e n_i) + \sum_i \delta n_i (\log_e g_i) = 0. \qquad ...(iii)$$

Let us now use the method of Lagrange's undetermined multipliers. We multiply eq. (i) and eq. (ii) by quantities - α and - β respectively (which do not depend upon any of the n_i's) and add the resulting expressions to eq. (iii). We get

$$-\sum_i \delta n_i(\log_e n_i)+\sum_i \delta n_i(\log_e g_i)-\sum_i \alpha\,\delta n_i-\sum_i \beta\,\varepsilon_i\,\delta n_i=0$$

or $$\sum_i(-\log_e n_i+\log_e g_i-\alpha-\beta\varepsilon_i)\delta n_i=0.$$

As δn_i's are independent variables, therefore, for the validity of the last equation, the quantity in parentheses must always be zero for each value of *i*. Hence

$$-\log_e n_i+\log_e g_i-\alpha-\beta\varepsilon_i=0.$$

or $$\boxed{n_i=g_i\,e^{-\alpha}\,e^{-\beta\varepsilon_i}.} \qquad \text{...(iv)}$$

This result determines the most probable distribution of molecules among the various energy states and is known as 'Maxwell-Boltzmann distribution law'. It expresses the general rule that *higher the state (higher the value of* ε_i*), the less is the population (less is the value of* n_i*).*

The Ensemble

In order to study a given system consisting of a large number of particles, we imagine a collection of a very large number of systems with the same macroscopic properties as the given system, but in different microscopic states. Such a hypothetical collection of "similar" systems is called an "ensemble". The members of the ensemble are known as "elements". Thus, the elements of an ensemble are identical in structure (same macroscopic state), but differ in the coordinates and momenta of the individual particles i.e. differ in their (unobservable) microscopic states*.

The concept of ensemble helps in understanding the actual system more completely. It represents at one time the (microscopic) states of the actual system which would occur in the course of time. It is easier to predict the statistical behaviour of a chosen ensemble than to study the behaviour of any particular system.

Ensemble Average : Suppose that a variable u of some system can take any of the n possible values $u_1, u_2, ...u_i, ... u_n$ with respective probabilities P(1), P(2), ...$P(i)$,..........$P(n)$. This means that in an ensemble of N similar systems, the variable u takes the particular value u_i in a number $N_i = N\,P\,(i)$ of these systems. Then, the ensemble average $\overline{\mu}$ is defined by

$$\overline{u} \equiv \frac{N_1 u_1 + N_2 u_2 + ... + N_i u_i + ... + N_n u_n}{N}$$

$$\equiv \frac{\sum_{i=1}^{n} N_i u_i}{N}$$

But, since $N_i / N = P(i)$ is the probability of occurrence of the value *ui, the* last expression may be written as

$$\overline{u} \equiv \sum_{i=1}^{n} P(i) u_i. \qquad ...(ii)$$

Similarly, if $f(u)$ is any function of u , the ensemble average (or mean value) of $f(u)$ is defined by

$$\overline{f}(u) \equiv \sum_{i=1}^{n} P(i) f(u_i). \qquad ...(iii)$$

Experimental measurements give only the average values for the ensemble, and not the values for the individual systems.

The ensemble averages have the following simple properties:

(i) *The ensemble average of a sum (or difference) of two functions is equal to the sum (or difference) of the ensemble averages of the two functions.*

If $f(u)$ and $g(u)$ are any two functions of u , then

$$\overline{f \pm g} = \overline{f} \pm \overline{g}. \qquad \text{...(iv)}$$

(ii) *The ensemble average of the product of a function and some constant is equal to the product of the ensemble average of the function and the constant.*

If c is any constant, then

$$\overline{cf} = c\overline{f}$$

(iii) *The ensemble average of a product of two functions is equal to the product of the ensemble averages of the two functions.*

If $f(u)$ and $f(v)$ are any functions of two statistically independent variables u and v , then

$$\overline{fg} = \overline{f}\,\overline{g}.$$

Fluctuations in Ensembles

With the passage of time the properties of a system fluctuate (vary) about the average of equilibrium values. The same is true about the elements of an ensemble.

Let us consider the property represented by the variable u. Its average value is $\overline{u}$. The deviation of u from its average value is defined by

$$\delta u = u - \overline{u}.$$

We note that the mean value of this deviation, $\overline{\delta u}$, is zero:

$$\overline{\delta u} = \overline{(u - \overline{u})}$$

$$= \overline{u} - \overline{u} \qquad \text{[by eq.(iv)]}$$

$$= 0$$

A rough measure of the fluctuation is provided by the 'mean-square deviation' :

$$\overline{(\delta u)^2} = \overline{(u - \bar{u})^2} = \overline{u^2} - 2\bar{u}\bar{u} + \left(\bar{u}\right)^2$$
$$= \overline{u^2} - \left(\bar{u}\right)^2$$

$\overline{(\delta u)^2}$ is called the "dispersion" of u. The *standard deviation* Δu of u is provided by the square-root of the dispersion, *i.e.* root-mean-square deviation from the average value. Thus

$$\Delta u = \left[\overline{(\delta u)^2}\right]^{1/2} = \left[\overline{u^2} - (\bar{u})^2\right]^{1/2}.$$

Energy Fluctuations in a Canonical Ensemble: When a system is in thermal equilibrium with a reservoir, the energy of the system fluctuates, because the energy is exchanged with the reservoir.

For a canonical ensemble, we have

$$\bar{E} = \frac{\sum_i E_i e^{-\beta E_i}}{\sum_i e^{-\beta E_i}} = -\frac{1}{Z}\frac{\partial Z}{\partial \beta}.$$

where $\quad Z = \sum_i e^{-\beta E_i}$ and $\beta = 1/kT$.

$$\therefore \qquad \overline{E^2} = \frac{\sum_i E_i^2 e^{-\beta E_i}}{\sum_i e^{-\beta E_i}} = \frac{1}{Z}\frac{\partial^2 Z}{\partial \beta^2}.$$

Now, differentiating the above first equation :

$$-\frac{\partial \bar{E}}{\partial \beta} = \frac{1}{Z}\left(\frac{\partial^2 Z}{\partial \beta^2}\right) - \frac{1}{Z^2}\left(\frac{\partial Z}{\partial \beta}\right)^2$$
$$= \overline{E^2} - (\bar{E})^2 = \overline{(\delta E)^2}$$

Now, the specific heat of a system at constant value of a parameter, say volume, is given by

$$C_V = \frac{\partial \bar{E}}{\partial T} = \frac{\partial \bar{E}}{\partial \beta}\frac{\partial \beta}{\partial T}$$

$$= \frac{\partial \bar{E}}{\partial \beta}\left(-\frac{1}{kT^2}\right) \qquad \left[\therefore \beta = \frac{1}{KT}, \therefore \frac{\partial \beta}{\partial \mathrm{T}} = -\frac{1}{KT^2}\right]$$

$$= \overline{(\delta E)^2}\,\frac{1}{KT^2}.$$

$$\therefore \overline{(\delta E)^2} = KT^2 C_v.$$

A measure of the energy fluctuation is the ratio

$$\frac{\delta E}{E} = \frac{\left[\overline{(\delta E)^2}\right]^{1/2}}{\bar{E}} = \frac{(KT^2\, C_V)^{1/2}}{\bar{E}}.$$

This is the expression for the fractional fluctuation in the energy of a cononical ensemble. It has a very small value for ordinary systems.

Let us consider the case of a perfect gas, for which we have.

$$C_V \approx Nk$$

and

$$\bar{E} \approx NkT.$$

Thus

$$\frac{\Delta E}{E} \approx \frac{(kT^2\, Nk)^{1/2}}{NkT} \approx \frac{1}{N^{1/2}}.$$

For 1 litre of perfect gas the number of molecules $N \approx 10^{22}$.

$$\therefore \frac{\Delta E}{E} \approx 10^{-11},$$

which is negligibly small.

PROBLEMS

1. What is the probability of drawing (i) a king, (ii) king of diamond, (iii) any card of diamond from a deck of 52 cards ?

Solution : (i) Total no. of cards = 52

no. of kings in the deck = 4.

$\therefore$ probability of getting a king in a draw

$$= \frac{4}{52} = \frac{1}{13}$$

(ii) There is only 1 king of diamond. Hence the probability of getting it

$$= \frac{1}{52}$$

(iii) There are 13 cards of diamond. Hence the probability of getting any card of diamond

$$= \frac{13}{52} = \frac{1}{4}.$$

2. A bag contains 6 green balls, 8 white balls and 10 black balls. If a ball is drawn from the bag, what is the probability of its being either white or black ?

Solution : The total number of balls is 6 + 8 + 10 = 24. The probability of getting a white ball is 8/24 , and that of getting a black ball is 10/24. Therefore, by addition theorem, the probability of getting either a white or a black ball is

$$\frac{8}{24} + \frac{10}{24} = \frac{18}{24} = \frac{3}{4}.$$

3. Find the probability that from two dice either 7 or 11 is obtained.

Solution : A die has six faces numbered 1, 2, 3, 4, 5, 6. From two dice there are 6 x 6 = 36 alternatives in all. The number of cases in which 7 is obtained are:

Die 1 :	1	2	3	4	5	6
Die 2 :	6	5	4	3	2	1

Thus, there are 6 cases in which the sum of two dice comes out to be 7. Hence the probability is

$$P(7) = \frac{6}{36}.$$

Similarly, there are only 2 cases (6, 5) and (5, 6) in which the sum comes out to be 11. Hence the corresponding probability is

$$P(11) = \frac{2}{36}.$$

Therefore, by addition theorem, the probability of getting either 7 or 11 is

$$P(7 \text{ or } 11) = P(7) + P(11) = \frac{6}{36} + \frac{2}{36} = \frac{2}{9}.$$

4. A bag contains 3 white and 4 black balls. If two balls are drawn out one after the other, what is the probability that the first ball will be white and the second ball will be black? The first ball is returned to the bag after it is taken out.

Solution : The total number of balls is 7. The total number of ways of drawing 2 balls from the bag is

$$^7C_2 = \frac{7!}{2!(7-2)!} = \frac{7 \times 6 \times 5!}{2 \times 5!} = 21.$$

The number of ways in which one white ball may be drawn out of the 3 is

$$^3C_1 = \frac{3!}{1!\,2!} = 3.$$

Similarly, the number of ways in which one black ball may be drawn out of the 4 is

$$^4C_1 = \frac{4!}{1!\,3!} = \frac{4 \times 3!}{1 \times 3!} = 4.$$

The probability of obtaining one white ball is

$$\frac{^3C_1}{^7C_2} = \frac{3}{21} = \frac{1}{7}.$$

and the probability of obtaining one black ball is

$$\frac{^4C_1}{^7C_2} = \frac{4}{21}.$$

By multiplication theorem, the probability of obtaining one white and one black ball is

$$\frac{1}{7} \times \frac{4}{21} = \frac{4}{147}.$$

5. From a pack of 52 cards two are drawn at random. Find the chance that one is a king and the other a queen.

Solution : The total number of ways of drawing 2 cards out of 52 is $^{52}C_2$.

There are 4 kings and 4 queens. Therefore, the number of ways of having a king and a queen is $^4C_1 \times {}^4C_1$.

$$\therefore \text{ required probability } = \frac{^4C_1 \times {}^4C_1}{^{52}C_2}$$

$$= \frac{\dfrac{4\,!}{1\,!\,3\,!} \times \dfrac{4\,!}{1\,!\,3\,!}}{\dfrac{52\,!}{2\,!\,50\,!}}$$

$$= \frac{4 \times 4}{\dfrac{52 \times 51}{2}} = \frac{8}{663}.$$

6. Find the probability that in tossing a coin 12 times we get (i) 4 heads, 8 tails, (ii) 6 heads, 6 tails.

Solution : The probability of getting r heads out of N tosses of a coin is given by

$$P(r) = \frac{N\,!}{r\,!\,(N-r)\,!}\, p^r q^{N-r},$$

where p is the probability of getting a 'head' and q is the probability of getting a 'tail' in a single toss.

In tossing a coin, there are only two outcomes, either a 'head' up or a 'tail' up and each is equally likely to occur. Thus, the probability of each is $\frac{1}{2}$ that is, $p = q = \frac{1}{2}$.

The last expression thus reduces to

$$P(r) = \frac{N\,!}{r\,!\,(N-r)\,!}\left(\frac{1}{2}\right)^N.$$

Here $N = 12$.

(i) The probability of getting 4 heads (r = 4) is

$$P(4) = \frac{12\,!}{4\,!\,(12-4)\,!}\left(\frac{1}{2}\right)^{12}$$

$$= \frac{12\,!}{4\,!\,8\,!}\left(\frac{1}{2}\right)^{12}$$

$$= \frac{(12 \times 11 \times 10 \times 9) \times 8\,!}{(4 \times 3 \times 2 \times 1) \times 8\,!}\left(\frac{1}{2}\right)^{12}$$

$$= 495\left(\frac{1}{2}\right)^{12} = \frac{495}{4096} = \mathbf{0.121}.$$

(ii) The probability of getting 6 heads ($r = 6$) is

$$P(6) = \frac{12\,!}{6\,!\,(12-6)\,!}\left(\frac{1}{2}\right)^{12}$$

$$= \frac{12\,!}{6\,!\,6\,!}\left(\frac{1}{2}\right)^{12}$$

$$= \frac{(12 \times 11 \times 10 \times 9 \times 8 \times 7) \times 6\,!}{(6 \times 5 \times 4 \times 3 \times 2 \times 1) \times 6\,!}\left(\frac{1}{2}\right)^{12}$$

$$= 924\left(\frac{1}{2}\right)^{12} = \frac{924}{4096} = \mathbf{0.225}$$

7. *Calculate the probability that in tossing a coin 10 times we get (i) all heads, (ii) 4 heads, 6 tails, (iii) 2 heads, 8 tails, (iv) 7 heads, 3 tails, (v) 1 head, 9 tails.* *(Agra 85)*

Solution : The probability of getting r heads out of N tosses of a coin is given by

$$P(r) = \frac{N\,!}{r\,!\,(N-r)\,!}\left(\frac{1}{2}\right)^N.$$

Here $N = 10$.

(i) The probability of getting all heads (r = 10) is

$$P(10) = \frac{10\,!}{10\,!\,(10-10)\,!}\left(\frac{1}{2}\right)^{10}$$

$$= 1\left(\frac{1}{2}\right)^{10} \qquad [\therefore\ 0\,! = 1]$$

$$= \frac{1}{1024} = 0.000976.$$

(ii) The probability of getting 4 heads ($r = 4$) is

$$P(4) = \frac{10\,!}{4\,!\,6\,!}\left(\frac{1}{2}\right)^{10}$$

$$= \frac{(10 \times 9 \times 8 \times 7) \times 6\,!}{(4 \times 3 \times 2 \times 1) \times 6\,!} \times \frac{1}{2^{10}}$$

$$= \frac{210}{2^{10}} = \frac{210}{1024}\ 0.205.$$

(iii) Similarly,

$$P(2) = \frac{10\,!}{2\,!\,8\,!}\left(\frac{1}{2}\right)^{10}$$

$$= \frac{(10 \times 9) \times 8\,!}{(2 \times 1) \times 8\,!} \times \frac{1}{2^{10}}$$

$$= \frac{45}{2^{10}} = \frac{45}{1024}\ 0.0439.$$

(iv) $$P(7) = \frac{10\,!}{7\,!\,3\,!}\left(\frac{1}{2}\right)^{10}$$

$$= \frac{(10 \times 9 \times 8) \times 7\,!}{7\,!\,(3 \times 2 \times 1)} \times \frac{1}{2^{10}}$$

$$= \frac{120}{2^{10}} = \frac{120}{1024} = 0.117.$$

(v) $$P(1)=\frac{10\,!}{1\,!\,9\,!}\left(\frac{1}{2}\right)^{10}$$

$$=\frac{10\times 9\,!}{1\times 9\,!}\left(\frac{1}{2}\right)^{10}$$

$$=\frac{10}{2^{10}}=\frac{10}{1024}\;0.00976.$$

8. Calculate the probability that in tossing a coin 8 times we get, (i) all heads, (ii) 4 heads, 4 tails, (iii) 3 heads, 5 tails, (iv) 5 heads, 3 tails.

(i) $(\frac{1}{2})^8$, (ii) $70\,(\frac{1}{2})^8$,

(iii) $56\,(\frac{1}{2})^8$, (iv) $56\,(\frac{1}{2})^8$

9. Eight similar coins are tossed simultaneously. Calculate (i) the probability of getting the heads of five coins uppermost, (ii) the probability of most probable combination, (iii) the probability of least probable combination.

Solution : The probability of getting r heads out of N tosses is

$$P(r)=\frac{N\,!}{r\,!\,(N-r)\,!}\,p^r\,q^{N-r},$$

where p is the probability of getting a 'head' uppermost and q is the probability of not getting it (or getting a 'tail' uppermost) in a single toss.

For a symmetrical coin, we have

$$p = q = \frac{1}{2}$$

$$\therefore\; P(r)=\frac{N\,!}{r\,!\,(N-r)\,!}\left(\frac{1}{2}\right)^N$$

Here $N = 8$.

(i) The probability of getting 5 heads ($r = 5$) is

$$P(5) = \frac{8\,!}{7\,!\,(8-5)\,!}\left(\frac{1}{2}\right)^8$$

$$= \frac{(8 \times 7 \times 6) \times 5\,!}{5\,! \times (3 \times 2 \times 1)}\left(\frac{1}{2}\right)^8$$

$$= 56\left(\frac{1}{2}\right)^8 = \frac{56}{256} = \frac{7}{32} = 0.2187\,.$$

(ii) The most probable combination is one for which

$$r = \frac{N}{2} = \frac{8}{2} = 4.$$

$$\therefore \quad P_{\max}(4) = \frac{8\,!}{4\,!\,(8-4)\,!}\left(\frac{1}{2}\right)^8$$

$$= \frac{(8 \times 7 \times 6 \times 5) \times 4\,!}{(4 \times 3 \times 2 \times 1) \times 4\,!}\left(\frac{1}{2}\right)^8$$

$$= 70\left(\frac{1}{2}\right)^8 = \frac{70}{256} = \frac{35}{128} = 0.2734\,.$$

(iii) The probability of least probable combination is

$$P_{\min} = \frac{1}{2^N} = \frac{1}{2^8} = \frac{1}{256} = 0.0039.$$

10. If 8 balls are distributed randomly between two boxes with equal probabilities, then calculate the probabilities of the most probable and least probable distributions.

$$\frac{70}{256}, \frac{1}{256}.$$

11. In a random distribution of 10 particles between two boxes with equal probabilities, calculate (i) probability of distribution (3,7), (ii) ratio of probability of this distribution with the distribution of largest probability, (iii) the total number of microstates and the number of macrostates and (iv) the number of microstates in macrostate (3,7).

Solution : In case of equal probabilities for 'two' states, the expression for the probability of distribution (r, $N - r$) is

$$P(r, N-r) = \frac{N!}{r!\,(N-r)!}\left(\frac{1}{2}\right)^N$$

(i) For $r = 3$, $N - r = 7$, $N = 10$, we have

$$P(3, 7) = \frac{10!}{3!\,7!}\left(\frac{1}{2}\right)^{10}$$

$$= \frac{(10 \times 9 \times 8) \times 7!}{(3 \times 2 \times 1) \times 7!}\left(\frac{1}{2}\right)^{10}$$

$$= 120 \times 2^{-10}$$

(ii) For the most probable distribution, $r = N/2 = 5$. Thus, the largest probability is given by

$$P(5,5) = \frac{10!}{5!\,5!}\left(\frac{1}{2}\right)^{10}$$

$$= \frac{(10 \times 9 \times 8 \times 7 \times 6) \times 5!}{(5 \times 4 \times 3 \times 2 \times 1) \times 5!}\left(\frac{1}{2}\right)^{10}$$

$$= 252 \times 2^{-10}$$

$$\therefore \quad \frac{P(3,7)}{P(5,5)} = \frac{120 \times 2^{-10}}{252 \times 2^{-10}} = \frac{10}{21}$$

(iii) Total number of microstates is $2^N = 2^{10}$

The total number of macrostates is the number of terms in the binomial expansion of $(a + b)^n$ *i.e.* $(a + b)^{10} = 11$.

(iv) The number of microstates in macrostate (3, 7)

$$= \frac{N!}{r!\,(N-r)!}$$

$$= \frac{10!}{3!\,7!} = 120.$$

12. A box has two halves of equal probabilities in which 14 particles are to be distributed. Find the probabilities of the distributions (10, 4); (14, 0); (7, 7).

$$1001(\tfrac{1}{2})^{14}, \quad (\tfrac{1}{2})^{14}, \quad 3432\ (\tfrac{1}{2})^{14}.$$

13. A coin is so loaded that the probability of getting head in a toss is 0 7. Deduce the probability that in 5 tosses we get (i) 2 heads, 3 tails ; (ii) all tails, (iii) all heads.

Solution : The probability of getting r heads out of N tosses is given by

$$P(r) = \frac{N\,!}{r\,!\,(N-r)\,!}\,p^r\,q^{N-r}$$

where p is the probability of getting 'head' up and q is the probability of not getting it or getting 'tail' up in a single toss. Thus, in a single toss

$$p + q = 1,$$

because the sum of the two probabilities is a certainty (either we shall get a 'head' up or a 'tail' up).

Here it is given that $p = 0.7$,

therefore $q = 1 - p = 1 - 0.7 = 0.3$.

Further, $N = 5$.

(i) The probability of getting 2 heads (and 3 tails) is

$$\begin{aligned} P(2) &= \frac{5\,!}{2\,!\,(5-2)\,!}\,(0.7)^2\,(0.3)^{5-2} \\ &= \frac{(5 \times 4) \times 3\,!}{(2 \times 1) \times 3\,!}\,(0.7)^2\,(0.3)^3 \\ &= 10(0.49)\,(0.027) \\ &= 0.1323. \end{aligned}$$

(ii) The probability of getting all tails (i.e. no head) is

$$P(0) = \frac{5\,!}{0\,!\,(5-0)\,!}\,(0.7)^0\,(0.3)^{5-0}$$

$$= 1\;(0.7)^0\,(0.3)^5$$

$$= 1(1)\,(0.00243)$$

$$= 0.00243.$$

(iii) The probability of getting all heads (r = 5) is

$$P(5) = \frac{5\,!}{5\,!\,(5-5)\,!}\,(0.7)^5\,(0.3)^{5-5}$$

$$= 1\;(0.7)^5\,(0.3)^0$$

$$= 1(0.7)^5\,(1)$$

$$= 0.1681.$$

14. There are two boxes with statistical weights 4 : 1, between which 10 particles are distributed. What is the probability for a distribution (2, 8) ? Which distribution has the maximum probability ?

Solution : The probability of distribution in which r particles go in one box and $(N - r)$ in the other box is

$$P(r, N-r) = \frac{N\,!}{r\,!\,(N-r)\,!}\,p^r\,q^{N-r},$$

where p is the probability of going of a particle in one box and q that of going of a particle in the other box, and

$$p + q = 1,$$

because the particle will go either in one box or in the other.

The statistical weights of the boxes are 4 : 1 *i.e.* probability for the first box is 4 times the probability for the other. Thus

$$p = 0.8 \text{ and } q = 0.2.$$

Hence for the distribution (2, 8) *i.e.* r = 2, $N - r$ = 8, the probability is

$$P(2,8) = \frac{10\,!}{2\,!\,8\,!}(0.8)^2(0.2)^8$$

$$= 45 \times (64 \times 10^{-2}) \times (256 \times 10^{-8})$$

$$= 7.37 \times 10^{-5}.$$

The most probable distribution is the one in which the particles are distributed in the ratio of the statistical weights (4 : 1) of the boxes, *i.e.* the distribution **(8, 2).**

15. Assume that each face of a 6-faced die is equally likely to land uppermost. Consider a game which involves the tossing of 5 such dice. Calculate the probability that the number "2" appears uppermost (i) in exactly one die, (ii) in exactly two dice, (iii) in atleast one die.

Solution : The probability of getting a given number (say "2") uppermost in tossing N dice is given by

$$P(r) = \frac{N\,!}{r\,!\,(N-r)\,!}\,p^r\,q^{N-r}$$

where r is number of dice showing "2" uppermost. Further, p is the probability of getting "2" uppermost in a single toss and q is the probability of not getting it.

Here $N = 5$. Since a die has 6 faces, the probability that any given die lands with "2" uppermost is 1/6, that is, $p = 1/6$. By symmetry, the probability that it does not land with "2" uppermost is $q = 1 - p = 1 - \frac{1}{6} = \frac{5}{6}$.

(i) The probability of getting "2" in exactly one die ($r = 1$) is

$$P(1) = \frac{5\,!}{1\,!\,(5-1)\,!}\left(\frac{1}{6}\right)^1\left(\frac{5}{6}\right)^{5-1}$$

$$= \frac{5 \times 4\,!}{4\,!}\left(\frac{1}{6}\right)\left(\frac{5}{6}\right)^4$$

$$= \left(\frac{5}{6}\right)^4 \simeq 0.4.$$

(ii) The probability of getting "2" in exactly two dice (r = 2) is

$$P(2) = \frac{5\,!}{2\,!(5-2)\,!}\left(\frac{1}{6}\right)^2\left(\frac{5}{6}\right)^{5-2}$$

$$= \frac{(5 \times 4) \times 3\,!}{(2 \times 1) \times 3\,!}\left(\frac{1}{6}\right)^2\left(\frac{5}{6}\right)^3$$

$$= \frac{1}{3}\left(\frac{5}{6}\right)^4 \simeq 0.16.$$

(iii) The probability of getting "2" in exactly one die is $(\frac{5}{6})^5$, as shown in (i). Hence the probability of getting "2" in atleast one die is $1 - (\frac{5}{6})^5 \simeq 1 - 0.4 \simeq 0.6$.

16. A six-faced die has its faces marked 1 to 6 and in a toss all the faces have equal probability of appearing up. If five such dice are tossed, calculate the probability of (i) all five dice showing up "4", (ii) just two dice showing up "4", (iii) none of the five dice showing up "4".

Solution : The probability of getting "4" up is given by

$$P(r) = \frac{N\,!}{r\,!\,(N-r)\,!}\,p^r\,q^{N-r}$$

where N is the total number of dice tossed, r is the number of dice showing up "4", p is the probability of getting up "4" in a single toss of one die and q is the probability of not getting "4" *(i.e.* $q = 1 - p$*)*.

Here N= 5, $p = 1/6$ and $q = 5/6$.

(i) The probability of all five dice showing up "4" (r = 5) is

$$P(5) = \frac{5\,!}{5\,!\,0\,!}\left(\frac{1}{6}\right)^5\left(\frac{5}{6}\right)^{5-5}$$

$$= \left(\frac{1}{6}\right)^5 \simeq 0.00013. \qquad [\because\ 0\,! = 1]$$

(ii) The probability of just two dice showing up '"4" (r = 2) is

$$P(2) = \frac{5\,!}{2\,!3\,!}\left(\frac{1}{6}\right)^2\left(\frac{5}{6}\right)^3$$

$$= \frac{(5 \times 4) \times 3\,!}{(2 \times 1) \times 3\,!}(5)^3\left(\frac{1}{6}\right)^5$$

$$= 2(5)^4\left(\frac{1}{6}\right)^5$$

$$= \frac{1}{3}\left(\frac{5}{6}\right)^4 \simeq 0.16.$$

(iii) The probability of none of the 5 dice showing up "4" (r = 0) is

$$P(0) = \frac{5\,!}{0\,!5\,!}\left(\frac{1}{6}\right)^0\left(\frac{5}{6}\right)^5$$

$$= \left(\frac{5}{6}\right)^5 \simeq 0.4.$$

17. Three six-faced dice are tossed simultaneously. The faces of each die are marked 1, 2, 3, 4, 5, 6. Find the probability of getting "1" uppermost of all the dice.

1/216 .

18. Consider 100 molecules and 10 cells of equal energy. All molecules are equally distributed among the cells. Calculate log, W, where W is thermodynamic probability.

Solution : The thermodynamic probability of distribution of N molecules among k cells, such that n_1 molecules go to cell 1, n_2 to cell 2, and so on is

$$W = \frac{N\,!}{n_1\,!\,n_2\,!\,......\,n_k\,!}.$$

Here $N = 100$, $k = 10$ and $n_1 = n_2 = n_3 = ... = n_{10}.$

$$\therefore \quad W = \frac{100\,!}{10\,!\;10\,!\;......\;\text{ten terms}}.$$

$$= \frac{100\,!}{(10!)^{10}}$$

Now, $\log_e W = \log_e 100\,! - 10 \log_e 10\,!$

By Stirling's approximation, $\log_e n\,! = (n \log_e n) - n$, we have

$$\log_e W \simeq \{(100 \log_e 100) - 100\} - 10\,\{(10 \log_e 10) - 10\}$$

$$\simeq 100 \log_e 100 - 100 \log_e 10$$

$$\simeq 100 \log_e (100/10) \simeq 100 \log_e 10$$

$$\simeq 100 \times 2.3 \simeq 230.$$

19. ***Consider an assembly of N = 13 identical particles having a fixed amount of total energy E = 20 joule. Let the particles be distributed among three cells 1, 2 and 3 such that $n_1 = 6$, $n_2 = 4$ and $n_3 = 3$ with energies $\varepsilon_1 = 0$, $\varepsilon_2 = 2$ joule per particle and $\varepsilon_3 = 4$ joule per particle. If $\delta n_3 = -2$, find δn_1, and δn_2, assuming N and E remaining constant. Also find the initial and final probabilities.***

Solution : The total number of particles is

$$N = n_1 + n_2 + n_3 = 6 + 4 + 3 = 13 \text{ (constant)}$$

and the total energy is

$$E = \varepsilon_1 n_1 + \varepsilon_2 n_2 + \varepsilon_3 n_3$$

$$= (0 \times 6) + (2 \times 4) + (4 \times 3) = 20 \text{ joule (constant).}$$

It is given that $\delta n_3 = -2$, that is, 2 particles leave the third (higher-energy) cell and go to the first and second (lower-energy) cells. Now,

$$n_1 + n_2 + n_3 = 13$$

so that

$$\delta n_1 + \delta n_2 + \delta n_3 = 0.$$

But $$\delta n_3 = -2.$$

$\therefore$ $$\delta n_1 + \delta n_2 = 2. \qquad ...(i)$$

Again, $$\varepsilon_1 n_1 + \varepsilon_2 n_2 + \varepsilon_3 n_3 = 20 \text{ joule}$$

so that

$$(0 \times n_1) + (2 \times n_2) + (4 \times n_3) = 20 \text{ joule}$$

or $$2n_2 + 4n_3 = 20 \text{ joule}$$

or $$2\,\delta n_2 + 4\,\delta n_3 = 0.$$

But $$\delta n_3 = -2.$$

$\therefore$ $$2\,\delta n_2 = 8$$

or $$\delta n_2 = 4.$$

Putting this value in eq. (i), we get

$$\delta n_1 = -2.$$

The new distribution of particles is :

$$n_1' = n_1 + \delta n_1 = 6 - 2 = 4.$$

$$n_2' = n_2 + \delta n_2 = 4 + 4 = 8.$$

$$n_3' = n_3 + \delta n_3 = 3 - 2 = 1.$$

$$N = n_1' + n_2' + n_3' = 4 + 8 + 1 = 13 \text{ (as before).}$$

Initial probability

$$= \frac{N\,!}{n_1\,!\,n_2\,!\,n_3\,!} = \frac{13\,!}{6\,!\,4\,!\,3\,!} = 60060.$$

Final probability

$$= \frac{N\,!}{n_1'\,!\,n_2'\,!\,n_3'\,!} = \frac{13\,!}{4\,!\,8\,!\,1\,!} = 6435.$$

20. A system has five different macrostates which have respectively 6, 20, 42, 12 and 2 accessible microstates. A certain property u of the system has values 4, 5, 2, 6 and 10 respectively for the five macrostates. Calculate (i) relative probabilities for the various macrostates, (ii) average values of u and u^2.

Solution: (i) The total number of microstates

$$= 6 + 20 + 42 + 12 + 2 = 82.$$

∴ relative probability for the five macrostates respectively are

$$\frac{6}{82}, \frac{20}{82}, \frac{42}{82}, \frac{12}{82}, \frac{2}{82}.$$

(ii) The average value of u is given by

$$\bar{u} = \sum_{i=1}^{5} P(i)\, u_i$$

$$= \left(\frac{6}{82} \times 4\right) + \left(\frac{20}{82} \times 5\right) + \left(\frac{42}{82} \times 2\right) + \left(\frac{12}{82} \times 6\right) + \left(\frac{2}{82} \times 10\right)$$

$$= \frac{24}{82} + \frac{100}{82} + \frac{84}{82} + \frac{72}{82} + \frac{20}{82}$$

$$= \frac{300}{82} = 3.66.$$

The average value of u^2 is given by

$$\overline{u^2} = \sum_{i=1}^{5} P(i)\, u_i^2$$

$$= \left(\frac{6}{82} \times 4^2\right) + \left(\frac{20}{82} \times 5^2\right) + \left(\frac{42}{82} \times 2^2\right) + \left(\frac{12}{82} \times 6^2\right) + \left(\frac{2}{82} \times 10^2\right)$$

$$= \frac{96}{82} + \frac{500}{82} + \frac{168}{82} + \frac{432}{82} + \frac{200}{82}$$

$$= \frac{1396}{82} = 17.$$

8

Perfect Differentials

Let us suppose that we have a certain quantity z depending upon two other quantities x and y, so that z is some single-valued function of x and y. Mathematically we express it as $z = f(x,y)$. Let us take mutually perpendicular axes of coordinates (Fig.). Then for any particular point of coordinates x, y, the quantity z has a particular and definite value. That is, when x and y are given, the quantity z is completely determined. The differential dz is then called "perfect differential". In this case, when we go from a point A to a point C, the value of z at C will be independent of the actual path adopted. There are a number of paths from A to C but the value of z changes in going from A and C by the same amount, whichever path is chosen. From this, it follows that if we take the quantity z through a cycle, such as $ABCD$, then

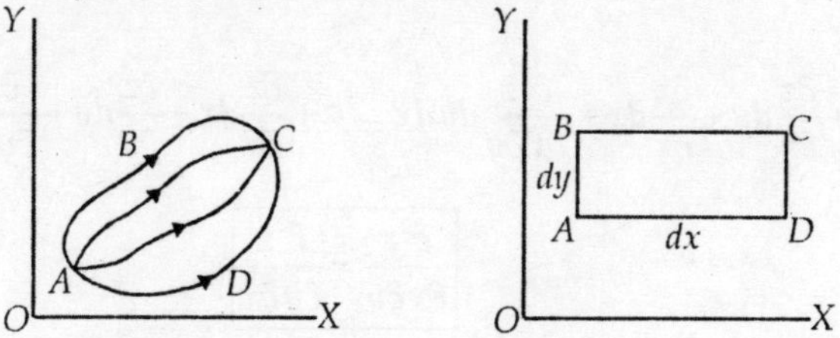

$$\oint dz = 0$$

Let us now find the mathematical condition for *dz* to be a perfect differential. Let the variables *x* and *y* change by infinitesimal amounts *dx* and *dy* respectively. To represent this, let us draw an elementary rectangle *ABCD*, the coordinates of *A* being *(x, y)* and the sides *AD* and *AB* of the rectangle being *dx* and *dy* respectively. Clearly, the coordinates of *D* will be $(x + dx, y)$; the coordinates of *B* will be $(x , y + dy)$, and the coordinates of *C* will be $(x + dx, y + dy)$.

Suppose, we choose the path *ABC* to go from *A* to *C*. Then if the value of the quantity is *z* at *A* , its value at *B* will be $z+\frac{\partial z}{\partial y}dy$. The value of z at C will be

$$\left(z+\frac{\partial z}{\partial y}dy\right)+\frac{\partial}{\partial x}\left(z+\frac{\partial z}{\partial y}dy\right)\times dx$$

Secondly, let us choose the path *ADC* from *A* to *C*. The value of *z* at D will be $z+\frac{\partial z}{\partial y}dx$. The value of z at C will be

$$\left(z+\frac{\partial z}{\partial x}dx\right)+\frac{\partial}{\partial y}\left(z+\frac{\partial z}{\partial y}dx\right)\times dy$$

Now, if *dz* is a perfect differential, the value of *z* must be the same at C , whichever path is chosen. That is

$$\left(z+\frac{\partial z}{\partial y}dy\right)+\frac{\partial}{\partial x}\left(z+\frac{\partial z}{\partial y}dy\right)dx = \left(z+\frac{\partial z}{\partial x}dx\right)+\frac{\partial}{\partial y}\left(z+\frac{\partial z}{\partial x}dx\right)dy$$

or $$z+\frac{\partial z}{\partial y}dy+\frac{\partial z}{\partial x}dx+\frac{\partial^2 z}{\partial x\,\partial y}dy\,dx = z+\frac{\partial z}{\partial x}dx+\frac{\partial z}{\partial y}dy+\frac{\partial^2 z}{\partial y\,\partial x}dx\,dy$$

or $$\boxed{\frac{\partial^2 z}{\partial x\,\partial y}=\frac{\partial^2 z}{\partial y\,\partial x}}$$

This is the mathematical condition for dz to be a perfect differential. If dz is a perfect differential of a function (x, y); it may be shown that

$$dz = \left(\frac{\partial z}{\partial x}\right)_y dx + \left(\frac{\partial z}{\partial y}\right)_x dy$$

Examples: The equilibrium state of a thermodynamic system is described in terms of variables p, V, T, U and S. All these five are state variables. If the system undergoes an infinitesimal change from one state to another, these variables change unfinitesimally. The differentials dp, dV, dT, dU and dS, which are involved in the changes, are all perfect differentials. Let us consider some cases :

Pressure : For 1 mole of an ideal gas, we have

$$pV = RT \qquad \text{where } R \text{ is a constant}$$

or

$$p = \frac{RT}{V}$$

Differentiating it with respect to T, taking V constant, we have

$$\frac{\partial p}{\partial T} = \frac{R}{V}$$

Again, differentiating with respect to V, we have

$$\frac{\partial}{\partial V}\left(\frac{\partial p}{\partial T}\right) = -\frac{R}{V^2} \qquad \ldots\text{(i)}$$

Let us now consider another series of changes. If, first the volume changes, temperature remaining constant, then

$$\frac{\partial p}{\partial V} = \frac{RT}{V^2}$$

If now the temperature changes, the volume, remaining constant, we have

$$\frac{\partial}{\partial T}\left(\frac{\partial p}{\partial V}\right) = -\frac{R}{V^2} \qquad \ldots\text{(ii)}$$

From eq. (i) and (ii), we have

$$\frac{\partial^2 p}{\partial V \partial T} = \frac{\partial^2 p}{\partial T \partial V}$$

Hence, *dp is a perfect differential*. Similarly, we can show that *dV and dT are perfect differentials.*

Internal Energy: When an amount of heat dQ is added to a substance, it is partly used up in increasing the internal energy of the substance by dU, and partly in doing the external work dW. From the first law of thermodynamics, we have

$$dQ = dU + dW.$$

In this equation , U, the internal energy, is completely determined for any state when we know the value of the coordinates which define that state. Therefore, the change in internal energy in passing from one state to another through a path is independent of the actual path chosen. *Hence dU is a perfect differential.* On the other hand, the external work done during the change depends not only on the initial and final conditions, but also on the path, and so dW is *not* a perfect differential. It follows from the above equation that dQ also is *not* a perfect differential.

Entropy : The second law of thermodynamics tells us that

$$dQ = T\,dS.$$

The entropy S is a property of the substance determined by the coordinates which define that particular state. In any reversible *cycle* $\oint dS$ *is zero and so dS is a perfect differential.*

We conclude that p ,V ,T, U and S are *point* functions, but Q and W are *path* functions.

Maxwell's Thermodynamic Relations

The state of a homogeneous system is completely determined if we know its mass and any two of the thermodynamic variables p , V ,T, U and S. Thus, the internal energy U of a system is completely determined if V and T are given *i.e.* U is a function of the two variables V and T. Among the five thermodynamic variables, certain relations exist, of which four are important and known as 'Maxwell's thermodynamic relations'. Let us deduce these relations.

According to the first and second laws of thermodynamics, we have respectively

$$dQ = dU + p\,dV \qquad [\because dW = p\,dV]$$

and $$dQ = T\,dS.$$

From these equations, we get

$$dU = T\,dS - p\,dV. \qquad \text{...(i)}$$

For the moment, let us call the two independent variables as x and y which determine the state of the system of a given mass. Now, any thermodynamic variable will be a function of x and y. Suppose U, S and V are functions of x and y . Then, we have

$$dU = \frac{\partial U}{\partial x}dx + \frac{\partial U}{\partial y}dy$$

$$dS = \frac{\partial S}{\partial x}dx + \frac{\partial S}{\partial y}dy$$

and $$dV = \frac{\partial V}{\partial x}dx + \frac{\partial V}{\partial y}dy$$

Putting these values of dU, dS and dV in eq. (i), we have

$$\frac{\partial U}{\partial x}dx + \frac{\partial U}{\partial y}dy = T\left(\frac{\partial S}{\partial x}dx + \frac{\partial S}{\partial y}dy\right) - P\left(\frac{\partial V}{\partial x}dx + \frac{\partial V}{\partial y}dy\right)$$

Equating the coefficients of dx and dy , we have

$$\frac{\partial U}{\partial x} = T\frac{\partial S}{\partial x} - P\frac{\partial V}{\partial x} \qquad \text{...(ii)}$$

and $$\frac{\partial U}{\partial y} = T\frac{\partial S}{\partial y} - P\frac{\partial V}{\partial y} \qquad \text{...(iii)}$$

Differentiating eq. (ii) with respect to y, and eq. (iii) with respect to x, we obtain

$$\frac{\partial^2 U}{\partial y\, dx} = \frac{\partial T}{\partial y}\frac{\partial S}{\partial x} + T\frac{\partial^2 S}{\partial x\,\partial y} - \frac{\partial p}{\partial y}\frac{\partial V}{\partial x} - p\frac{\partial^2 V}{\partial y\,\partial x}$$

and $$\frac{\partial^2 U}{\partial x\,\partial y} = \frac{\partial T}{\partial x}\frac{\partial S}{\partial y} + T\frac{\partial^2 S}{\partial x\,\partial y} - \frac{\partial p}{\partial x}\frac{\partial V}{\partial y} - p\frac{\partial^2 V}{\partial x\,\partial y}$$

But $$\frac{\partial^2 U}{\partial y\,\partial x} = \frac{\partial^2 U}{\partial x\,\partial y}, \frac{\partial^2 S}{\partial y\,\partial x} = \frac{\partial^2 S}{\partial x\,\partial y} \text{ and } \frac{\partial^2 V}{\partial y\,\partial x} = \frac{\partial^2 V}{\partial x\,\partial y},$$

since dU, dS and dV are all "perfect differentials." Hence, we get

$$\frac{\partial T}{\partial y}\frac{\partial S}{\partial x} - \frac{\partial p}{\partial y}\frac{\partial v}{\partial x} = \frac{\partial T}{\partial x}\frac{\partial S}{\partial y} - \frac{\partial p}{\partial x}\frac{\partial V}{\partial y} \qquad \text{...(iv)}$$

This is the general Maxwell's thermodynamic equation. Let us now substitute any pair of the variables p , V, T and S for the independent variables x and y .

Let us put x = S and y = V, we have then

$$\frac{\partial S}{\partial x} = 1; \; \frac{\partial V}{\partial y} = 1,$$

and $$\frac{\partial S}{\partial y} = 0; \; \frac{\partial V}{\partial x} = 0.$$

Therefore , equation (iv) gives

$$\frac{\partial T}{\partial y} = -\frac{\partial p}{\partial x}$$

But $\partial y = \partial$ V (as y = V)

and $\partial x = \partial$ S (as x = 5).

Therefore

$$\frac{\partial T}{\partial V} = -\frac{\partial p}{\partial S}$$

or $$\boxed{\left(\frac{\partial T}{\partial V}\right)_S = -\left(\frac{\partial p}{\partial S}\right)_V} \qquad \ldots(1)$$

Similarly, putting x = T and y = V, we get from eq. (iv)

$$\boxed{\left(\frac{\partial S}{\partial V}\right)_T = \left(\frac{\partial p}{\partial T}\right)_V} \qquad \ldots(2)$$

Putting x = S and y = p , we get

$$\boxed{\left(\frac{\partial T}{\partial p}\right)_S = \left(\frac{\partial V}{\partial S}\right)_p} \qquad \ldots(3)$$

Finally, putting x = T and y = p , we get

$$\boxed{\left(\frac{\partial S}{\partial p}\right)_T = \left(\frac{\partial V}{\partial S}\right)_p} \qquad \ldots(4)$$

Equations (1) to (4) are known as *Maxwell's four thermodynamic relations.*

Prove that :

(i) $$\left(\frac{\partial p}{\partial S}\right)_T \left(\frac{\partial S}{\partial T}\right)_p = -\left(\frac{\partial p}{\partial T}\right)_S$$

(ii) $$\left(\frac{\partial p}{\partial V}\right)_T \left(\frac{\partial V}{\partial T}\right)_p = -\left(\frac{\partial p}{\partial T}\right)_V$$

From the fourth Maxwell's relation

$$\left(\frac{\partial S}{\partial p}\right)_T = -\left(\frac{\partial V}{\partial T}\right)_p$$

we have $\left(\frac{\partial p}{\partial S}\right)_T = -\left(\frac{\partial T}{\partial V}\right)_p$

$$\therefore \quad \left(\frac{\partial p}{\partial S}\right)_T\left(\frac{\partial S}{\partial T}\right)_p = -\left(\frac{\partial T}{\partial V}\right)_p\left(\frac{\partial S}{\partial T}\right) = -\left(\frac{\partial S}{\partial V}\right)_p \qquad \text{...(i)}$$

From the third relation $\left(\frac{\partial T}{\partial p}\right)_S = \left(\frac{\partial V}{\partial S}\right)_p$, we have

$$\left(\frac{\partial S}{\partial V}\right)_p = \left(\frac{\partial p}{\partial T}\right)_S$$

Making this substitution in eq. (i), we get

$$\left(\frac{\partial p}{\partial S}\right)_T\left(\frac{\partial S}{\partial T}\right)_p = -\left(\frac{\partial p}{\partial T}\right)_S$$

This is the required relation.

(ii) From the fourth relation, $\left(\frac{\partial V}{\partial T}\right)_p = -\left(\frac{\partial S}{\partial p}\right)_T$

$$\therefore \quad \left(\frac{\partial p}{\partial V}\right)_T\left(\frac{\partial V}{\partial T}\right)_p = -\left(\frac{\partial p}{\partial V}\right)_T\left(\frac{\partial S}{\partial p}\right)_T = -\left(\frac{\partial S}{\partial V}\right)_T$$

But $\left(\frac{\partial S}{\partial V}\right)_T = \left(\frac{\partial p}{\partial T}\right)_V$ by second relation.

$$\therefore \quad \left(\frac{\partial p}{\partial V}\right)_T\left(\frac{\partial V}{\partial T}\right)_p = -\left(\frac{\partial p}{\partial T}\right)_V$$

Prove the thermodynamic relation

$$\left(\frac{\partial S}{\partial V}\right)_T = \left(\frac{\partial p}{\partial T}\right)_V$$

and hence prove that $\frac{dp}{dT} = \frac{L}{T(V_2 - V_1)}$

Explain the phenomenon of regulation of ice by the above relation.

The general expression for Maxwell's thermodynamic relation is

$$\frac{\partial T}{\partial y}\frac{\partial S}{\partial x}-\frac{\partial p}{\partial y}\frac{\partial V}{\partial x} = \frac{\partial T}{\partial x}\frac{\partial S}{\partial y}-\frac{\partial p}{\partial x}\frac{\partial V}{\partial y}$$

Let us put $x = T$ and $y = V$. We have then

$$\frac{\partial T}{\partial x}=1, \qquad \frac{\partial V}{\partial y}=1,$$

and $$\frac{\partial T}{\partial y}=0, \qquad \frac{\partial V}{\partial x}=0,$$

Therefore, the general expression written above gives

$$\frac{\partial S}{\partial y}-\frac{\partial p}{\partial x} = 0$$

But $\partial y = \partial V$ (as y = V) and $\partial x = \partial T$ (as x = T) . Therefore

$$\left(\frac{\partial S}{\partial V}\right)_T = \left(\frac{\partial p}{\partial T}\right)_V$$

This is the required relation.

Multiplying both sides by T, we get

$$T\left(\frac{\partial S}{\partial V}\right)_T = T\left(\frac{\partial p}{\partial T}\right)_V$$

But T dS = dQ . Therefore

$$\left(\frac{\partial Q}{\partial V}\right)_T = \left(\frac{\partial p}{\partial T}\right)_V$$

Here dQ represents the heat absorbed *at constant temperature*. Therefore , dQ is the latent heat L. Thus, if a body absorbs an amount of heat L (latent heat) to change its state at constant temperature T, and the specific volumes in the first and the second states are V_1 and V_2, we obtain from the last expression

$$\frac{L}{V_2 - V_1} = T\frac{dp}{dT}$$

or
$$\frac{dp}{dT} = \frac{L}{T(V_2 - V_1)}$$

This is Clausius-Clapeyron latent heat equation.

Regulation of Ice

'If two pieces of ice are pressed together, a single piece is formed on releasing the pressure. This phenomenon is known as regelation.' It can be explained from the last expression.

We know that ice contracts on melting *i.e.* V_2 is less than V_1 or (V_2- V_1)) is negative. Therefore, dp/dT is also negative which means that *melting point of ice is lowered with increase in pressure.* Thus , when two pieces of ice are pressed together, the melting point is lowered so that some ice melts and a thin film of water is formed between the two surfaces in contact. On removing the pressure, the original melting point is restored. Hence the water film is frozen, thereby cementing together the two ice pieces.

(a) Prove the thermodynamic relation :

$$\mathbf{TdS} = C_v dT + T\left(\frac{\partial p}{\partial T}\right)_V dV$$

Let the entropy S of a thermodynamic system be a function of temperature T and volume V, *i.e.*

$$S = f(T,V).$$

Since dS is a perfect differential, we can write

$$dS = \left(\frac{\partial S}{\partial T}\right)_V dT + \left(\frac{\partial S}{\partial V}\right)_T dV.$$

Multiplying by T; we get

$$TdS = T\left(\frac{\partial S}{\partial T}\right)_V dT + T\left(\frac{\partial S}{\partial V}\right)_T dV. \qquad \text{...(i)}$$

For any substance, the specific heat at constant volume is given by

$$C_V = \left(\frac{\partial Q}{\partial T}\right)_V = T\left(\frac{\partial S}{\partial T}\right)_V \qquad [\because dQ = TdS]$$

and $$\left(\frac{\partial S}{\partial V}\right)_T = \left(\frac{\partial p}{\partial T}\right)_V. \qquad \text{[Maxwell's (2) relation]}$$

Making these two substitutions in eq. (i), we get

$$\boxed{TdS = C_V dT + T\left(\frac{\partial p}{\partial T}\right)_V dV.} \qquad \ldots\text{(ii)}$$

This is known as 'first $T\,dS$ equation'.

(b) One mole of a Van der Waals' gas suffers reversible isothermal expansion from a volume V_1 to V_2. How much heat has been transferred ?

Heat Transfer in Isothermal Expansion : The Van der Waals eq. of state for 1 mole of gas is

$$\left(p + \frac{q}{V^2}\right)(V - b) = \text{RT}$$

or $$p = \frac{RT}{V-b} - \frac{a}{V^2}$$

or $$\left(\frac{\partial p}{\partial T}\right)_V = \frac{R}{V-b}$$

Substituting this in the first TdS equation (ii), we get

$$TdS = C_V dT + RT\frac{dV}{V-b}$$

Integrating, we get

$$\int TdS = C_V \int dT + RT\int \frac{dV}{V-b}$$

Let ΔQ be the heat transferred when the gas undergoes reversible 'isothermal' expansion ($dT = 0$) from V_1 to V_2. Then

$$\Delta Q = \int T dS = RT \int_{V_1}^{V_2} \frac{dV}{V-b}$$

$$= RT\left[\log_e (V-b)\right]_{V_1}^{V_2}$$

$$= RT \log_e \frac{V_2 - b}{V_1 - b}$$

If R is given in terms of erg or joule then the value of ΔQ will also be obtained in the same units. Then, we shall write

$$\Delta Q = \frac{RT}{J} \log_e \frac{V_2 - b}{V_1 - b} \, cal$$

(a) Prove the thermodynamic relation :

$$TdS = C_p dT - T\left(\frac{\partial V}{\partial T}\right)_P dp$$

Let the entropy S of a thermodynamic system be a function of temperature T and pressure p, *i.e.*

$$S = f(T,p).$$

Since dS is a perfect differential, we can write

$$dS = \left(\frac{\partial S}{\partial T}\right)_p dT + \left(\frac{\partial S}{\partial p}\right)_T dp$$

Multiplying by T; we get

$$T\,ds = T\left(\frac{\partial S}{\partial T}\right)_P dT + T\left(\frac{\partial S}{\partial P}\right)_T dp.$$

But $T\left(\frac{\partial S}{\partial T}\right)_p = \left(\frac{\partial Q}{\partial T}\right)_p = C_p$, the specific heat at constant pressure, and $\left(\frac{\partial S}{\partial p}\right)_p = -\left(\frac{\partial V}{\partial T}\right)_p$, Maxwell's fourth relation.

$$\therefore \quad \boxed{TdS = C_p dT - T\left(\frac{\partial V}{\partial T}\right)_p dp.} \qquad \text{...(i)}$$

This is known as 'second TdS equation'.

(b) A liquid is compressed isothermally. How much heat would be transferred? What would have been the change in temperature if the liquid had been compressed adiabatically?

Heat Transfer in Isothermal Compression of Liquid: Let ΔQ be the heat transferred when the liquid is compressed isothermally from a pressure p_1 to p_2. Then using the last relation, we have

$$\Delta Q \int TdS = C_p \int dT - T \int_{P_1}^{P_2} \left(\frac{\partial V}{\partial T}\right)_p dp$$

Since the compression is isothermal *(dT = O)*, we have

$$\Delta Q = -T \int_{P_1}^{P_2} \left(\frac{\partial V}{\partial T}\right)_p dp$$

Now, $\frac{1}{V}\left(\frac{\partial V}{\partial T}\right)_p$ represents increase in volume per unit volume per unit increase in temperature at constant pressure. Thus, it represents coefficient of volume expansion α.

$$\therefore \quad \left(\frac{\partial V}{\partial T}\right)_p = V\alpha$$

Putting this value of $\left(\frac{\partial V}{\partial T}\right)_p$ in the last expression, we get

$$\Delta Q = -T \int_{P_1}^{P_2} V\alpha\, dp$$

In case of a liquid (and solid), V and α are independent (practically) of pressure.

Hence, we may write

$$\Delta Q = -TV\alpha \int_{p_1}^{p_2} dp = -TV\alpha(p_2 - p_1)$$

$$= -\frac{TV\alpha}{J}(p_2 - p_1)\text{Cal}$$

The negative sign is significant. This shows that as the pressure is increased isothermally, heat will flow *out* of the liquid if its coefficient of expansion α is positive ; and absorbed by the liquid from outside if its coefficient of expansion is negative (such as water between 0 and 4°C).

Adiabatic Compression

If the liquid had been compressed adiabatically, the heat evolved or absorbed would cause a change in the internal energy and hence a change in temperature. If m be the mass and s the specific heat of the liquid, the change in temperature is given by

$$\Delta T = \frac{\Delta Q}{ms} = \frac{TV\alpha}{Jms}(p_2 - p_1)$$

Let $\frac{V}{m} = 0$ (the specific volume). Then

$$\Delta T = \frac{Tv\alpha}{Js}(p_2 - p_1).$$

This shows that an adiabatic increase in pressure will heat a substance with a positive coefficient of expansion α, and cool a substance with a negative α.

An important relation for a thermodynamic system is the second TdS equation which is as under :

$$TdS = C_p dT - T\left(\frac{\partial V}{\partial T}\right)_p dp$$

Let us treat the wire as a thermodynamic system. Let h be the length, and dl the increase in length when stretched isothermally by a stretching force (negative pressure) F. The work done is $-F\,dl$ (instead of $p\,dV$ in a gas). Hence $\left(\frac{\partial V}{\partial T}\right)_p dp$ in the above equation must be replaced by $\left(\frac{\partial l}{\partial T}\right)_F dF$. Further, let s be the specific heat of the material of the wire which is independent of pressure. Then C_p in the above equation must be replaced by s . Then the above equation becomes

$$TdS = sdT + T\left(\frac{\partial l}{\partial T}\right)_F dF$$

Let ΔQ be the heat transferred when the stretching force is increased from F_1 to F_2. Then

$$\Delta Q = \int TdS = s\int dT + T\int_{F_1}^{F_2}\left(\frac{\partial l}{\partial T}\right)_F dF$$

Since the stretching is isothermal ($dT = 0$), we have

Now $\frac{1}{l}\left(\frac{\partial l}{\partial T}\right)_F$ represents the increase in length per unit length per unit increase in temperature under a constant stretching. Thus it represents the coefficient of linear expansion α .

$$\therefore \quad \left(\frac{\partial l}{\partial T}\right)_F = l\alpha$$

Putting this value of $(\partial l/\partial T)_F$ in the last expression, we get

$$\Delta Q = Tl\alpha\int_{F_1}^{F_2} dF$$

$$= Tl\alpha(F_2 - F_1) = Tl\alpha\Delta F$$

$$= \frac{Tl\alpha}{J}\Delta F \text{ cal.}$$

This is the expression for the heat transferred.

If the wire be stretched adiabatically, the necessary work will come from the internal energy which would cause a change in temperature. If m be the mass and s the specific heat of the wire, the change in temperature is given by

$$\Delta T = \frac{\Delta Q}{ms}$$

$$= \frac{Tl\alpha}{Jms}\Delta F$$

Let $\frac{m}{l} = w$, the mass per unit length of the wire.

$$\therefore \quad \Delta T = \frac{T\alpha}{Jws}\Delta F$$

This is the expression for the change in temperature. An increase in tension (ΔF) means a decrease in pressure, so that when α is positive (for all metals α is positive), the wire will be *cooled* when stretched adiabatically.

Rubber contracts on heating *i.e.* it has a negative coefficient of expansion α, so that ΔT" is positive. This means that a rubber string shows a heating effect when stretched adiabatically.

Let a thermodynamic system undergo a reversible change from one equilibrium state to another, its behaviour may be represented by a set of thermodynamic relations. For example, the Maxwell's fourth relation is

$$\left(\frac{\partial S}{\partial p}\right)_T = -\left(\frac{\partial V}{\partial T}\right)_P$$

Multiplying both sides by T, we get

$$T\left(\frac{\partial S}{\partial p}\right)_T = -T\left(\frac{\partial V}{\partial T}\right)_P$$

But $T\partial S = \partial Q$.

$$\therefore \qquad \left(\frac{\partial Q}{\partial p}\right)_T = -T\left(\frac{\partial V}{\partial T}\right)_P$$

This is the required relation. Further $\frac{1}{V}\left(\frac{\partial V}{\partial T}\right)_P$ is the increase in volume per unit volume per unit increase in temperature at a constant pressure. Thus it represents the coefficient of volume expansion α.

$$\therefore \qquad \left(\frac{\partial V}{\partial T}\right)_P = V\alpha$$

Putting this value of $\left(\frac{\partial V}{\partial T}\right)_P$ in the last expression, we have

$$\left(\frac{\partial Q}{\partial p}\right)_T = -TV\alpha$$

This is the required relation. It shows that if α is positive, *i.e.* if the substance expands on heating, then $\left(\frac{\partial Q}{\partial p}\right)_T$ will be negative. That is, the heat must be withdrawn from the body when the pressure is increased, if the temperature is to be kept constant. In other words, the increase in pressure will heat a body which expands on heating. On the other hand, if a is negative *i. e.* if the body contracts on heating, then $\left(\frac{\partial Q}{\partial p}\right)_T$ will be positive. That is, the heat must be added to the body when the pressure is increased, if the temperature is to be kept constant. In other words, the increase in pressure will cool a body which contracts on heating.

Prove the following :

(a) $\left(\frac{\partial C_p}{\partial p}\right)_T = -T\left(\frac{\partial^2 V}{\partial T^2}\right)_P$

(b) $\left(\frac{\partial C_V}{\partial V}\right)_T = T\left(\frac{\partial^2 p}{\partial T^2}\right)_V$

(a) We start with Maxwell's fourth relation :

$$\left(\frac{\partial S}{\partial p}\right)_T = -\left(\frac{\partial V}{\partial T}\right)_P$$

Differentiating partially with respect to T at constant pressure, we have

$$\left\{\frac{\partial}{\partial T}\left(\frac{\partial S}{\partial p}\right)_T\right\}_P = -\left(\frac{\partial^2 V}{\partial T^2}\right)_P$$

Multiplying both sides by T:

$$T\left\{\frac{\partial}{\partial T}\left(\frac{\partial S}{\partial p}\right)_T\right\}_P = -T\left(\frac{\partial^2 V}{\partial T^2}\right)_P$$

or $$\left\{T\frac{\partial}{\partial p}\left(\frac{\partial Q}{\partial T}\right)_T\right\}_P = -T\left(\frac{\partial^2 V}{\partial T^2}\right)_P$$

But $T\,dS = dQ$. Therefore

$$\left\{\frac{\partial}{\partial p}\left(\frac{\partial S}{\partial T}\right)_P\right\}_T = -T\left(\frac{\partial^2 V}{\partial T^2}\right)$$

Now, $\left(\frac{\partial Q}{\partial T}\right)_P = C_P$. Therefore

$$\left(\frac{\partial C_P}{\partial p}\right)_T = -T\left(\frac{\partial^2 V}{\partial T^2}\right)_P$$

(b) Let us start with Maxwell's second relation :

$$\left(\frac{\partial S}{\partial V}\right)_T = \left(\frac{\partial p}{\partial T}\right)_V$$

Differentiating with respect to T at constant volume, we have

$$\text{or} \quad \left\{\frac{\partial}{\partial T}\left(\frac{\partial S}{\partial V}\right)_T\right\}_V = \left(\frac{\partial^2 p}{\partial T^2}\right)_V$$

$$\text{or} \quad T\left\{\frac{\partial}{\partial T}\left(\frac{\partial S}{\partial V}\right)_T\right\}_V = T\left(\frac{\partial^2 p}{\partial T^2}\right)_V$$

$$\text{or} \quad T\left\{\frac{\partial}{\partial V}\left(\frac{\partial S}{\partial T}\right)_V\right\}_T = T\left(\frac{\partial^2 p}{\partial T^2}\right)_V$$

$$\text{or} \quad \left\{\frac{\partial}{\partial V}\left(\frac{\partial Q}{\partial T}\right)_V\right\}_T = T\left(\frac{\partial^2 p}{\partial T^2}\right)_V$$

$$\text{or} \quad \left(\frac{\partial C_V}{\partial V}\right)_T = T\left(\frac{\partial^2 p}{\partial T^2}\right)_V$$

(a) Prove from the laws of thermodynamics

$$\left(\frac{\partial U}{\partial V}\right)_T = T\left(\frac{\partial p}{\partial T}\right)_V - P$$

Hence show that the internal energy of an ideal gas is independent of its volume, but not so is the internal energy of an actual gas.

(b) Prove for a Van der Waals' gas :

$$U = \int_0^T C_V dT - \frac{a}{V} + U_0$$

Hence show that the internal energy of an actual gas increases as the volume increases at constant temperature.

(a) *Energy Equation :* Let a thermodynamic system undergo an infinitesimal reversible process between two equilibrium states. Its general behaviour may be expressed by the first and second laws of thermodynamics, which are

$$dQ = dU + pdV$$

and $$dQ = TdS.$$

These give

$$dU + pdV = TdS$$

or $$dS = \frac{dU + pdV}{T}$$

Making this substitution in maxwell's 2nd relation $\left(\frac{\partial S}{\partial V}\right)_T = \left(\frac{\partial p}{\partial T}\right)_V$ we get

$$\frac{1}{T}\left(\frac{\partial U + p\partial V}{\partial V}\right)_T = \left(\frac{\partial p}{\partial \mathrm{T}}\right)_V$$

or $$\left(\frac{\partial U}{\partial V}\right)_T + P = T\left(\frac{\partial p}{\partial T}\right)_V$$

$$\therefore \quad \boxed{\left(\frac{\partial U}{\partial V}\right)_T = T\left(\frac{\partial p}{\partial T}\right)_V - p} \qquad \text{...(i)}$$

This is the required energy equation. Let us apply it to gases :

(i) *Ideal Gas :* For 1 mole of an ideal gas, we have

$$p = \frac{RT}{V}$$

or $$\left(\frac{\partial p}{\partial T}\right)_V = \frac{R}{V}$$

or $$T\left(\frac{\partial p}{\partial T}\right)_V = \frac{RT}{V} = p$$

Putting this in eq. (i), we get

$$\left(\frac{\partial U}{\partial V}\right)_T = 0$$

Thus , the internal energy of an ideal gas is independent of its volume at constant temperature.

(ii) *Van der Waals' Gas* : For 1 mole of an actual gas, we have

$$\left(p+\frac{a}{V^2}\right)(V-b) = RT \qquad \text{...(ii)}$$

or
$$p = \frac{RT}{V-b}-\frac{a}{V^2}$$

or
$$\left(\frac{\partial p}{\partial T}\right)_V = \frac{R}{V-b}$$

Putting this in eq. (i), we get

$$\left(\frac{\partial U}{\partial V}\right)_T = \frac{RT}{V-b}-p$$

Putting the value of $\frac{RT}{V-b}-p$ from eq. (ii), we have

$$\left(\frac{\partial U}{\partial V}\right)_T = \frac{a}{V^2}$$

which is positive. This means that *the internal energy of an actual gas does depend on the volume at constant temperature.*

(b) For a Van der Waals' gas, we have

$$\left(\frac{\partial U}{\partial V}\right)_T = \frac{a}{V^2}$$

Let dU be a perfect differential of T and V. Then , we have

$$dU = \left(\frac{\partial U}{\partial T}\right)_V dT+\left(\frac{\partial U}{\partial V}\right)_T dV$$

Now,
$$\left(\frac{\partial U}{\partial T}\right)_V = \left(\frac{\partial Q}{\partial T}\right)_V = C_V$$

because at constant volume *($dW = pdV = 0$)*; we have $dQ = dU$ by first law of thermodynamics,

and
$$\left(\frac{\partial U}{\partial V}\right)_T = \frac{a}{V^2}$$

$$\therefore \qquad dU = C_V dT + \frac{a}{V^2} dV$$

Integrating:

$$U = \int C_V dT = -\frac{a}{V} + U_0 \qquad \text{(constant)}$$

This means that *the internal energy of an actual gas increases as the volume increases at constant temperature.*

Joule's Law

This law states that the internal energy of a perfect gas is independent of the volume of the gas at constant temperature:

$$\left(\frac{\partial U}{\partial V}\right)_T = 0$$

Van der Waals' Gas under Free Expansion : Suppose 1 mole of a real gas undergoes *free* expansion. Let ΔT be the change in temperature corresponding to change in volume ΔV. Then , by reciprocity theorem, we have

$$\left(\frac{\partial T}{\partial V}\right)_u = -\left(\frac{\partial T}{\partial U}\right)_V \left(\frac{\partial U}{\partial V}\right)_T \qquad \ldots\text{(i)}$$

Now, the molar specific heat at constant volume is defined as

$$C_V = \left(\frac{\partial Q}{\partial T}\right)_V$$

By first law : $\quad dQ = dU + pdV$

$$= dU. \qquad \text{(at constant volume)}$$

$$\therefore \qquad C_V = \left(\frac{\partial U}{\partial T}\right)_V$$

Making this substitution in eq. (i), we get

$$\left(\frac{\partial T}{\partial V}\right)_U = -\frac{1}{C_V}\left(\frac{\partial U}{\partial V}\right)_T$$

We have proved in the last question that for a Van der Waals' gas, we have

$$\left(\frac{\partial U}{\partial V}\right)_V = \frac{a}{V^2}$$

$$\therefore \qquad \left(\frac{\partial T}{\partial V}\right)_V = -\frac{1}{C_V}\frac{a}{V^2}$$

For a finite change in volume, the change in temperature is

$$\Delta T = -\frac{1}{C_V}\frac{a}{V^2}\Delta V$$

The Maxwell's four thermodynamic relations for a homogeneous body are as follows :

$$\left(\frac{\partial T}{\partial V}\right)_S = -\left(\frac{\partial p}{\partial S}\right)_V \qquad \text{...(i)}$$

$$\left(\frac{\partial S}{\partial V}\right)_T = \left(\frac{\partial p}{\partial T}\right)_V \qquad \text{...(ii)}$$

$$\left(\frac{\partial T}{\partial p}\right)_S = \left(\frac{\partial V}{\partial S}\right)_P \qquad \text{...(iii)}$$

$$\left(\frac{\partial S}{\partial p}\right)_T = -\left(\frac{\partial V}{\partial T}\right)_P \qquad \text{...(iv)}$$

The coefficient of volume elasticity E is defined as

$$E = -\frac{dp}{dV/V} = -V\frac{dp}{dV}$$

The isothermal elasticity E_T (temperature constant) and the adiabatic elasticity E_S (entropy constant) may be expressed by

$$E_T = V\left(\frac{\partial p}{\partial V}\right)_T$$

and
$$E_S = V\left(\frac{\partial p}{\partial V}\right)_S$$

$$\therefore \quad \frac{E_S}{E_T} = \frac{\left(\frac{\partial p}{\partial V}\right)_S}{\left(\frac{\partial p}{\partial V}\right)_T} = \frac{\left(\frac{\partial p}{\partial T}\frac{\partial T}{\partial V}\right)_S}{\left(\frac{\partial p}{\partial S}\frac{\partial S}{\partial V}\right)_T} = \frac{\left(\frac{\partial p}{\partial T}\right)_S\left(\frac{\partial T}{\partial V}\right)_S}{\left(\frac{\partial p}{\partial S}\right)_T\left(\frac{\partial S}{\partial V}\right)_T} \quad \text{...(i)}$$

But
$$\left(\frac{\partial p}{\partial T}\right)_S = \left(\frac{\partial S}{\partial V}\right)_p \quad \text{from (iii)}$$

$$\left(\frac{\partial T}{\partial V}\right)_S = -\left(\frac{\partial p}{\partial S}\right)_V \quad \text{from (i)}$$

$$\left(\frac{\partial p}{\partial S}\right)_T = -\left(\frac{\partial T}{\partial S}\right)_V \quad \text{from (iv)}$$

and
$$\left(\frac{\partial S}{\partial V}\right)_T = \left(\frac{\partial p}{\partial T}\right)_V \quad \text{from (ii)}$$

Putting these values in (i), we get

$$\frac{E_S}{E_T} = \frac{\left(\frac{\partial S}{\partial V}\right)_P\left(\frac{\partial p}{\partial S}\right)_V}{\left(\frac{\partial T}{\partial V}\right)_P\left(\frac{\partial p}{\partial T}\right)_V} = \left(\frac{\partial S/\partial V}{\partial T/\partial V}\right)_P\left(\frac{\partial p/\partial S}{\partial p/\partial T}\right)_V$$

$$= \left(\frac{\partial S}{\partial T}\right)_P\left(\frac{\partial T}{\partial S}\right)_V = \frac{\left(\frac{\partial S}{\partial T}\right)_P}{\left(\frac{\partial S}{\partial T}\right)_V}$$

Multiplying the numerator and denominator by T, we get

$$\frac{E_S}{E_T} = \frac{T\left(\frac{\partial S}{\partial T}\right)_P}{T\left(\frac{\partial S}{\partial T}\right)_V}$$

But $TdS = dQ$

$$\therefore \qquad \frac{E_S}{E_T} = \frac{\left(\frac{\partial Q}{\partial T}\right)_P}{\left(\frac{\partial Q}{\partial T}\right)_V}$$

$$\text{Now} \quad \left(\frac{\partial Q}{\partial T}\right)_P = C_p \text{ and } \left(\frac{\partial Q}{\partial T}\right)_V = C_V$$

$$\therefore \qquad \frac{E_S}{E_T} = \frac{C_p}{C_V}$$

Prove that for a homogeneous fluid

$$C_p - C_V = T\left(\frac{\partial p}{\partial T}\right)_V \left(\frac{\partial V}{\partial T}\right)_P$$

Hence Prove that for a perfect gas $C_p - CV = R$ ***and for a Van der Waals' gas***

$$C_P - C_V = R\left(1 + \frac{2a}{RTV}\right)$$

the symbols have their usual meanings.

Or

Establish the relation

$$C_P - C_V = \left[p + \left(\frac{\partial U}{\partial V}\right)_T\right]\left(\frac{\partial V}{\partial T}\right)_P$$

The molar specific heats at constant pressure and at constant volume of a thermodynamic system are expressed by

$$C_p = \left(\frac{\partial Q}{\partial T}\right)_P \text{ and } C_V = \left(\frac{\partial Q}{\partial T}\right)_V$$

$$\therefore \qquad C_P - C_V = \left(\frac{\partial Q}{\partial T}\right)_P - \left(\frac{\partial Q}{\partial T}\right)_V$$

But $dQ = TdS$.

$$\therefore \quad C_P - C_V = T\left(\frac{\partial S}{\partial T}\right)_P - T\left(\frac{\partial S}{\partial T}\right)_V \qquad \text{...(i)}$$

Now, if the entropy S is considered as a function of T and V, then , since dS is a perfect differential, we have

$$dS = \left(\frac{\partial S}{\partial T}\right)_p \partial T + \left(\frac{\partial S}{\partial T}\right)_T dV$$

or

$$\left(\frac{\partial S}{\partial T}\right)_p = \left(\frac{\partial S}{\partial T}\right)_V + \left(\frac{\partial S}{\partial V}\right)_T \left(\frac{\partial V}{\partial T}\right)_P$$

Multiplying by T throughout, we have

$$T\left(\frac{\partial S}{\partial T}\right)_P - T\left(\frac{\partial S}{\partial T}\right)_V = T\left(\frac{\partial S}{\partial V}\right)_T \left(\frac{\partial V}{\partial T}\right)_P$$

Substituting this in eq. (i), we have

$$C_P - C_V = T\left(\frac{\partial S}{\partial V}\right)_T \left(\frac{\partial V}{\partial T}\right)_P \qquad \text{...(ii)}$$

But $\left(\frac{\partial S}{\partial V}\right)_T = \left(\frac{\partial p}{\partial T}\right)_V$ by Maxwell's (2) relation.

$$\therefore \quad \boxed{C_p - C_V = T\left(\frac{\partial p}{\partial T}\right)_V \left(\frac{\partial V}{\partial T}\right)_P} \qquad \text{...(iii)}$$

This is the required expression.

Now, from the first and second laws, we can write

$$TdS = dU + pdV$$

or

$$T\left(\frac{\partial S}{\partial V}\right)_T = \left(\frac{\partial U}{\partial V}\right)_T + p$$

Making this substitution in eq. (ii), we have

$$\boxed{C_p - C_V = \left[p + \left(\frac{\partial U}{\partial V}\right)_T\right]\left(\frac{\partial V}{\partial T}\right)_T} \quad \ldots(iv)$$

This is the alternative expression.

Perfect Gas : For one mole of a perfect gas, we have

$$pV = RT$$

$$\therefore \quad \left(\frac{\partial p}{\partial T}\right)_p = \frac{R}{V}$$

and $$\left(\frac{\partial V}{\partial T}\right)_p = \frac{R}{p}$$

Making these substitutions in eq. (iii), we get

$$C_p - C_V = \frac{R^2T}{pV}$$

But $pV = RT$.

$$\therefore \quad C_p - C_v = R.$$

Van der Waals' Gas : For a Van der Waals' gas, we have

$$\left(p + \frac{a}{V^2}\right)(V - b) = \text{RT}$$

where a and b are constants. This may be written as

$$\left(p + \frac{a}{V^2}\right) = \frac{RT}{V - b}$$

Differentiating it with respect to Tat constant volume, we get

$$\left(\frac{\partial p}{\partial T}\right)_V = \frac{RT}{V-b}$$

and differentiating with respect to T at constant pressure,

$$-\frac{2a}{V^3}\left(\frac{\partial V}{\partial T}\right)_P = \frac{RT}{(V-b)^2}\left(\frac{\partial V}{\partial T}\right)_P + \frac{R}{V-b}$$

or $$\left(\frac{\partial V}{\partial T}\right)_P\left[\frac{RT}{(V-b)^2} - \frac{2a}{V^3}\right] = \frac{R}{V-b}$$

or $$\left(\frac{\partial V}{\partial T}\right)_P = \frac{\dfrac{R}{V-b}}{\dfrac{RT}{(V-b)^2} - \dfrac{2a}{V^3}}$$

Substituting the values of $\left(\frac{\partial p}{\partial T}\right)_V$ and $\left(\frac{\partial V}{\partial T}\right)_T$ in eq. (iii), we get

$$C_p - C_V = \frac{T\dfrac{R}{(V-b)}\dfrac{R}{(V-b)}}{\dfrac{RT}{(V-b)^2} - \dfrac{2a}{V^3}} = \frac{R}{1-\left(\dfrac{2a}{V^3}\right)\left(\dfrac{V-b}{RT}\right)^2}$$

Since $b << V$; we may replace (V - b) by V in the second power. Then , we get

$$C_p - C_V = \frac{R}{1-\left(\dfrac{2a}{V^3}\right)\dfrac{V^2}{RT}}$$

$$= R\left(1-\frac{2a}{RTV}\right)^{-1}$$

Expanding by binomial theorem and remembering that a is small, we get

$$C_p - C_V = R\left(1+\frac{2a}{RTV}\right)$$

Prove from first principles the relation

$$C_p - C_V = T\left(\frac{\partial p}{\partial T}\right)_V \left(\frac{\partial V}{\partial T}\right)_P = TE\alpha^2 V$$

where C_p and C_v are the molar specific heats at constant pressure and at constant volume respectively, E is the bulk modulus of elasticity; α is the coefficient of volume expansion and V is the specific volume. Discuss its significance.

The molar specific heats at constant pressure and at constant volume of a thermodynamic system are expressed by

$$C_P = \left(\frac{\partial Q}{\partial T}\right)_P \text{ and } C_V = \left(\frac{\partial Q}{\partial T}\right)_V$$

$$\therefore \quad C_p - C_V = \left(\frac{\partial Q}{\partial T}\right)_P \left(\frac{\partial Q}{\partial T}\right)_V$$

But $dQ = TdS$.

$$\therefore \quad C_p - C_V = T\left(\frac{\partial S}{\partial T}\right)_P - T\left(\frac{\partial S}{\partial T}\right)_V \quad \text{...(i)}$$

Now, if the entropy S is considered as a function of T and V, then since dS is a perfect differential, we have

$$dS = \left(\frac{\partial S}{\partial T}\right)_V dT + \left(\frac{\partial S}{\partial V}\right)_T dV$$

or

$$\left(\frac{\partial S}{\partial T}\right)_P = \left(\frac{\partial S}{\partial T}\right)_V + \left(\frac{\partial S}{\partial V}\right)_T \left(\frac{\partial V}{\partial T}\right)_P$$

Multiplying by T throughout, we have

$$T\left(\frac{\partial S}{\partial T}\right)_P - T\left(\frac{\partial S}{\partial T}\right)_V = T\left(\frac{\partial S}{\partial V}\right)_T \left(\frac{\partial V}{\partial T}\right)_P$$

Substituting this in eq. (i), we have

$$C_p - C_V = T\left(\frac{\partial S}{\partial V}\right)_T \left(\frac{\partial V}{\partial T}\right)_P$$

But $\left(\frac{\partial S}{\partial V}\right)_T = \left(\frac{\partial p}{\partial T}\right)_V$ by Maxwell's (2) relation.

$$\therefore \qquad \boxed{C_p - C_V = T\left(\frac{\partial p}{\partial T}\right)_V \left(\frac{\partial V}{\partial T}\right)_p} \qquad \ldots\text{(ii)}$$

Now, since p is a function of T and V, and dp is a perfect differential, we have

$$dp = \left(\frac{\partial p}{\partial T}\right)_V dT + \left(\frac{\partial p}{\partial V}\right)_T dV.$$

If the change takes place under constant pressure, $dp = 0$. Then

$$\left(\frac{\partial p}{\partial T}\right)_V dT = -\left(\frac{\partial p}{\partial V}\right)_T dV$$

or

$$\left(\frac{\partial p}{\partial T}\right)_V = -\left(\frac{\partial p}{\partial V}\right)_T \left(\frac{\partial V}{\partial T}\right)_p$$

Putting this value of $\left(\frac{\partial p}{\partial T}\right)_p$ in eq. (ii), we have

$$\boxed{C_p - C_V = T\left(\frac{\partial p}{\partial V}\right)_T \left(\frac{\partial V}{\partial T}\right)_p^2}$$

$$= -\frac{T}{V}\left(\frac{\partial p}{\partial V/V}\right)_T V^2 \left(\frac{\partial V/V}{\partial T}\right)_p^2$$

$$= -T\left(\frac{\partial p}{\partial V/V}\right)_T \left(\frac{\partial V/V}{\partial T}\right)_p^2 V$$

But $-\left(\frac{\partial p}{\partial V/V}\right)_T = E,$ the bulk modulus at constant temperature and $\left(\frac{\partial V/V}{\partial T}\right)_P = \alpha$, the coefficient of volume expansion. Therefore

$$\boxed{C_p - C_V = TE\alpha^2 V}$$

We can draw the following conclusions from this equation:

(i) The bulk modulus E, as defined by $-\left(\frac{\partial p}{\partial V/V}\right)_T$, is always positive for all known substances, and α^2 must be positive. Thus, $C_p - C_v$ can never be negative *i.e.* C_p can never be less than C_v.

(ii) As $T \to 0, C_p \to C_V$, *i.e.* at the absolute zero the two specific heats are equal.

The adiabatic (entropy constant) and isobaric (pressure constant) coefficients of volume expansion are defined as

$$\alpha_S = \frac{1}{V}\left(\frac{\partial V}{\partial T}\right)_S$$

and
$$\alpha_P = \frac{1}{V}\left(\frac{\partial V}{\partial T}\right)_P$$

$$\therefore \qquad \frac{\alpha_S}{\alpha_p} = \frac{\left(\frac{\partial V}{\partial T}\right)_S}{\left(\frac{\partial T}{\partial T}\right)_P} = \frac{1}{\left(\frac{\partial T}{\partial V}\right)_S \left(\frac{\partial V}{\partial T}\right)_P}$$

But $\left(\frac{\partial T}{\partial V}\right)_S = -\left(\frac{\partial p}{\partial S}\right)_V$, by Maxwell's first relation.

$$\therefore \qquad \frac{\alpha_s}{\alpha_p} = -\frac{1}{\left(\frac{\partial p}{\partial S}\right)_V \left(\frac{\partial V}{\partial T}\right)_P}$$

$$= -\frac{1}{\left(\frac{\partial p}{\partial T}\right)_V \left(\frac{\partial T}{\partial S}\right)_V \left(\frac{\partial V}{\partial T}\right)_P}$$

$$= -\frac{\left(\frac{\partial S}{\partial T}\right)_V}{\left(\frac{\partial p}{\partial T}\right)_V \left(\frac{\partial V}{\partial T}\right)_P}$$

$$= \frac{T\left(\frac{\partial S}{\partial T}\right)_P}{T\left(\frac{\partial p}{\partial T}\right)_V \left(\frac{\partial V}{\partial T}\right)_p} = -\frac{\left(\frac{\partial Q}{\partial T}\right)_P}{T\left(\frac{\partial p}{\partial T}\right)_V \left(\frac{\partial V}{\partial T}\right)_P}$$

But $T\left(\frac{\partial p}{\partial T}\right)_V \left(\frac{\partial V}{\partial T}\right)_P = C_P - C_V$

and $\left(\frac{\partial Q}{\partial T}\right)_V = C_V$.

$$\therefore \quad \frac{\alpha_S}{\alpha_p} = \frac{C_V}{C_P - C_V} = -\frac{1}{\frac{C_p}{C_V} - 1} = \frac{1}{1-\gamma}$$

The adiabatic (entropy constant) and isochoric (volume Constant) pressure coefficients are define as

$$\beta_S = \frac{1}{p}\left(\frac{\partial p}{\partial T}\right)_S$$

and

$$\beta_V = \frac{1}{P}\left(\frac{\partial P}{\partial T}\right)_V$$

$$\therefore \frac{\beta_S}{\beta_V} = \frac{\left(\frac{\partial p}{\partial T}\right)_S}{\left(\frac{\partial P}{\partial T}\right)_V} = \frac{1}{\left(\frac{\partial T}{\partial P}\right)_S \left(\frac{\partial P}{\partial T}\right)_V}$$

But $\left(\frac{\partial T}{\partial P}\right)_S = \left(\frac{\partial V}{\partial S}\right)_P$, by Maxwell's third relation.

$$\therefore \frac{\beta_S}{\beta_V} = \frac{1}{\left(\frac{\partial V}{\partial S}\right)_P \left(\frac{\partial P}{\partial T}\right)_V}$$

$$\frac{1}{\left(\frac{\partial V}{\partial T}\right)_P \left(\frac{\partial T}{\partial S}\right)_P \left(\frac{\partial p}{\partial T}\right)_V}$$

$$= \frac{\left(\frac{\partial S}{\partial T}\right)_P}{\left(\frac{\partial V}{\partial T}\right)_P \left(\frac{\partial P}{\partial T}\right)_V}$$

$$= \frac{T\left(\frac{\partial S}{\partial T}\right)_P}{T\left(\frac{\partial V}{\partial T}\right)_P \left(\frac{\partial P}{\partial T}\right)_V}$$

$$= -\frac{\left(\frac{\partial Q}{\partial T}\right)}{T\left(\frac{\partial V}{\partial T}\right)_P \left(\frac{\partial P}{\partial T}\right)_V}$$

$$= \frac{C_p}{C_r - C_V}$$

$$= \frac{C_p / C_V}{(C_P / C_V) - 1} = \frac{\gamma}{\gamma - 1}$$

Joule-Thomson Effect

When a gas under a constant pressure is made to pass through an insulated porous-plug to a region of lower constant pressure, it suffers a change in temperature. This is called the "Joule-Thomson or Joule-Kelvin effect". The process is called the "throttling process". The change in temperature is proportional to the pressure-difference between the two sides of the plug. At ordinary temperatures, all gases, except hydrogen and helium, show a cooling effect while hydrogen and helium show a small heating effect.

Expression for Cooling : Let us consider 1 mole of a gas. Let p_1, and V_1, be its pressure and volume before passing, and p_2 and V_2 the pressure and volume after passing through the porous plug. The *net* external work done by the gas in passing through the plug is then $p_2V_2 - p_1V_1$. Since there is no heat-exchange between the gas and its surroundings, this work must come from the internal energy of the gas. Thus, U_1 and U_2 be the internal energies of the gas before and after passing through the plug, we have from the first law of thermodynamics

$$U_1 - U_2 = p_2V_2 - p_1V_1$$

or $$U_1 + p_1V_1 = U_2 + p_2V_2$$

or $$\boxed{U + pV = \text{Constant}}$$

The quantity $(U + pV)$ which remains constant during a throttling process, is called the *enthalpy (H)* of the gas. Thus, we may write

$$dH = d(U + pV) = 0$$

or $$dU + pdV + Vdp = 0$$

But $dU + pdV = dQ$ (first law of thermodynamics) and $dQ = TdS$ (second law of thermodynamics). Therefore

$$TdS + Vdp = 0. \qquad ...(i)$$

Let us regard the entropy S as the function of variables p and T. Since dS is a perfect differential, we have

$$ds = \left(\frac{\partial S}{\partial T}\right)_P dt + \left(\frac{\partial S}{\partial p}\right)_T dp.$$

Substituting this value of dS in eq. (i), we get

But $TdS = dQ$. Therefore, $T\left(\frac{\partial S}{\partial T}\right)_P = \left(\frac{\partial Q}{\partial T}\right)_P = JC_P$ where C_p is the molar sp. heat at constant pressure in cal/(mole-°C) and $\left(\frac{\partial S}{\partial P}\right)_T = -\left(\frac{\partial V}{\partial T}\right)_P$ by Maxwell's fourth relation. Therefore

$$JC_p dT - T\left(\frac{\partial V}{\partial T}\right)_p dp + Vdp = 0$$

or $$JC_p dT = \left[T\left(\frac{\partial V}{\partial T}\right)_p - V\right]dp$$

or $$\frac{dT}{dp} = \frac{1}{JC_p}\left[T\left(\frac{\partial V}{\partial T}\right)_p - V\right]$$

Since the enthalpy H remains constant during the throttling process, we write $\frac{dT}{dp}$ as $\left(\frac{\partial T}{\partial P}\right)_H$ which is called the Joule-Thomson coefficient μ. Thus

$$\boxed{\mu = \left(\frac{\partial T}{\partial P}\right)_H = \frac{1}{JC_p}\left[T\left(\frac{\partial V}{\partial T}\right) - V\right]} \qquad \text{...(ii)}$$

Integrating it, we get the temperature-change for a finite drop in pressure from p_1 to p_2 as

$$\Delta T = \frac{1}{JC_p}\int_{P_1}^{P_2}\left[T\left(\frac{\partial V}{\partial T}\right)_P - V\right]dp \qquad \text{...(iii)}$$

Let us apply it to a perfect gas and an actual gas.

Perfect Gas : For 1 mole of a perfect gas, the eq. of state is $pV - RT$. Differentiating it with respect to T, taking p constant, we have

$$\left(\frac{\partial V}{\partial T}\right)_P = \frac{R}{p}$$

or $$T\left(\frac{\partial V}{\partial T}\right)_P = \frac{R}{p}T = V \qquad [\because pV = RT]$$

or $$T\left(\frac{\partial V}{\partial T}\right)_P - V = 0$$

Substituting this result in eq. (ii) and (iii), we get,

$$\boxed{\mu = \Delta T = 0}$$

Thus , *the Joule-Thomson effect for a perfect gas is zero.*

Van der Waals' Gas : For a Van der Waals' gas, we have

$$\left(p+\frac{a}{V^2}\right)(V-b) \;=\; RT$$

Differentiating it with respect to T, taking p constant, we have

$$\left(p+\frac{a}{V^2}\right)\left(\frac{\partial V}{\partial T}\right)_p - \frac{2a}{V^3}\left(\frac{\partial V}{\partial T}\right)_p (V-b) \;=\; R$$

or
$$\left(\frac{\partial V}{\partial T}\right)_p \;=\; \frac{R}{\left(p+\dfrac{a}{V^2}\right)-\dfrac{2a(V-b)}{V^3}}$$

$$=\; \frac{R}{\dfrac{RT}{V-b}-\dfrac{2a}{V^3}(V-b)} = \frac{R(V-b)}{RT-\dfrac{2a}{V^3}(V-b)^2}$$

$$\therefore \quad T\left(\frac{\partial V}{\partial T}\right)_p - V \;=\; \frac{RT(V-b)}{RT-\dfrac{2a}{V^3}(V-b)^2} - V$$

$$=\; \frac{-RTb+\dfrac{2a}{V^3}(V-b)^2}{RT-\dfrac{2a}{V^3}(V-b)^2}$$

$$=\; \frac{2aV(V-b)^2 - RTV^3 b}{RTV^3 - 2a(V-b)^2} \qquad \ldots\text{(iv)}$$

Since a and b are very small quantities, we replace $2aV\,(V-b)^2$ by $2aV^3$ in the numerator and ignore $2a\,(V-b)^2$ in comparison with RTV^3 in the denominator. Then, we get

$$T\left(\frac{\partial V}{\partial T}\right)_p - V = \frac{2aV^3 - RTV^3b}{RTV^3} = \frac{2a}{RT} - b$$

$$= \frac{2a}{RT} - b$$

Substituting this result in eq. (ii) and (iii), we get

$$\boxed{\mu = \frac{1}{JC_p}\left(\frac{2a}{RT} - b\right)}$$

and $$\Delta T = \frac{1}{JC_p}\int_{p_1}^{p_2}\left(\frac{2a}{RT} - b\right)dp$$

$$= \frac{\frac{2a}{RT} - b}{JC_p}(p_2 - p_1)$$

The pressure is lower on the emergent side of the porous plug ($p_2 < p_1$). Hence we may write as

$$\Delta T = \left(\frac{\frac{2a}{RT} - b}{JC_p}\right)(p_2 - p_1)$$

This expression shows that if $\frac{2a}{RT} > b$ or $T < \frac{2a}{RT}$ then ΔT will be negative *i.e.* the gas will be *cooled* on passing through the porous plug. If $T > \frac{2a}{Rb}$, then ΔT will be positive and the gas will be *warmed*. At $T = \frac{2a}{Rb}$, we have $\Delta T = 0$ *i.e.* there is no change in the temperature of the gas on passing through the porous plug. This is called the 'temperature of inversion' T;. At this temperature the Joule-Thomson coefficient $\mu = 0$.

$$\boxed{T_i = \frac{2a}{Rb} = \frac{27}{4}T_c = 2T_B}$$

For most gases the temperature of inversion is greater than the ordinary temperatures. Hence a cooling effect is obtained. For hydrogen and helium, however, the inversion temperatures are much below the ordinary temperatures, hence they show a heating effect at ordinary temperatures. If, however, they are pre-cooled below their temperatures of inversion they will also show a cooling effect.

An accurate expression for inversion temperature may be obtained by equating the exact value of $T\left(\frac{\partial V}{\partial T}\right)_p - V$, given by eq. (iv), to zero. Then

$$2aV(V-b)^2 - RT_iV^3b = 0$$

$$\therefore \qquad T_i = \frac{2a(V-b)^2}{RV^2b}.$$

Joule-Thomson Effect as due to deviation from Joule's Law and Boyle's Law : The Joule-Thomson effect, as deduced in the last question, is expressed as

$$\left(\frac{\partial T}{\partial p}\right)_H = \frac{1}{JC_p}\left[T\left(\frac{\partial V}{\partial T}\right)_p - V\right]$$

By Maxwell's fourth relation : $\left(\frac{\partial V}{\partial T}\right)_p = -\left(\frac{\partial S}{\partial p}\right)_T \cdot$ Thus

$$\left(\frac{\partial T}{\partial p}\right)_H = -\frac{1}{JC_P}\left[T\left(\frac{\partial S}{\partial P}\right)_T + V\right].$$

Now. $T\,dS = dQ = dU + p\,dV$(thermodynamical laws.) Therefore

$$\left(\frac{\partial T}{\partial p}\right)_H = \frac{1}{JC_p}\left[T\left(\frac{\partial U}{\partial p}\right)_T + \left(\frac{p\partial V}{\partial p}\right)_T + V\right]$$

$$\text{or} \qquad JC_p\left(\frac{\partial T}{\partial p}\right)_H = -\left(\frac{\partial U}{\partial p}\right)_T - \left[\frac{\partial (pV)}{\partial p}\right]_T$$

According to Joule's law, U_T is constant, while according to Boyle's law $(pV)_T$ is constant. Thus , the first term on the right-hand-side represents deviation from the Joule's law while the second term represents deviation from the Boyle's law. The Joule-Thomson effect is thus the resultant of these two deviations.

Cooling by Adiabatic Demagnetisation : When a paramagnetic solid is placed in a magnetising field *H*, its elementary magnetic dipoles become aligned parallel to the field. The magnetic moment per unit volume thus produced is called the 'intensity of magnetisation' *I*. The work spent in magnetising the solid appears as heat and, if conditions are adiabatic, the temperature of the solid rises. Conversely, an adiabatic demagnetisation produces a drop in temperature. This phenomenon is known as 'magneto-caloric effect.'

Experiments show that the intensity of magnetisation *I* produced by a magnetising field *H* in a paramagnetic solid at Kelvin temperature *T* is proportional to *H/T*, so that

$$I = C(H/T),$$

where C is known as 'Curie's constant'. For I mole of a paramagnetic substance, the (molar) intensity of magnetisation would be

$$M = IV = (CV)H/T, \qquad \text{...(i)}$$

where *V* is the molar volume (volume of 1 mole). *CV* is the Curie's constant for I mole.

Let us now consider 1 mole of a paramagnetic solid placed in a magnetising field *H* . The thermodynamic behaviour of a chemical system is expressed in terms of the variables p, V, S and *T*.

Here (- *H*) may be taken to play a role similar to that of *p* , and *M* that of *V* . Hence the Maxwell's third thermodynamic relation $\left(\frac{\partial T}{\partial p}\right)_S = \left(\frac{\partial V}{\partial S}\right)_p$ will hold provided we replace p by $-H$ and V by M. That is

$$\left(\frac{\partial T}{\partial H}\right)_S = -\left(\frac{\partial M}{\partial S}\right)_H$$

or

$$\left(\frac{\partial T}{\partial H}\right)_S = \frac{\left(\frac{\partial M}{\partial T}\right)_H}{\left(\frac{\partial S}{\partial T}\right)_H} = -\frac{T\left(\frac{\partial M}{\partial T}\right)_H}{T\left(\frac{\partial S}{\partial T}\right)_H}$$

$$= -\frac{T\left(\frac{\partial M}{\partial T}\right)_H}{\left(\frac{\partial Q}{\partial T}\right)_H} \qquad [\because T\,dS = dQ]$$

But $\left(\frac{\partial Q}{\partial T}\right)_H = JC_H$, where C_H *is* the specific heat of the substance in cal/(mole-deg) under constant field.

$$\therefore \quad \left(\frac{\partial T}{\partial H}\right)_S = -\frac{T}{JC_H}\left(\frac{\partial M}{\partial T}\right)_H$$

For an adiabatic (S constant) change in the field, we may write it as

$$dT = -\frac{T}{JC_H}\left(\frac{\partial M}{\partial T}\right) dH$$

Integrating it, we get the temperature change for a finite change in field from H_1 to H_2 as

$$\Delta T = -\frac{T}{JC_H}\int_{H_1}^{H_2}\left(\frac{\partial M}{\partial T}\right)_H dH$$

We have regarded both T and C_H as constant (an approximation) because an adiabatic change in field produces a temperature-change, which is only a small fraction of the initial temperature.

Now, from eq. (i), $M = (CV)\ H/T$ *so* that $\left(\frac{\partial M}{\partial T}\right)_H = -\frac{CVH}{T^2}$.

Therefore, the last expression becomes

$$\Delta T = \frac{T}{JC_H}\left(-\frac{CV}{T^2}\right)\int_{H_1}^{H_2} H\,dH$$

$$= \frac{CV}{2JC_H}T\left(H_2^2 - H_1^2\right)$$

Suppose the field is reduced from a value H to zero, *i.e.* $H_1 = H$ and $H_2 = 0$, then the change in temperature is given by

$$\Delta T = \frac{CV}{2JC_H T}H^2$$

The negative sign indicates that as the field is reduced, the temperature drops. Thus, *greater is the initial field and lower the initial temperature, greater is the temperature-drop.* CV is the Curie constant per mole. If 1 gm of the paramagnetic substance is taken, then CV would stand for Curie constant per gm.

The most important application of the magneto-caloric effect is in the production of extremely low temperatures, those below 1 K.

PROBLEMS

1. The temperature of a gas (coefficient of volume expansion 0.0037 per K and isothermal bulk modulus of elasticity 10^5 dynes/cm^2), which is not a perfect gas, is changed by 0.1 K at constant volume. Calculate the change in pressure.

Solution : Let p, V, T be the state variables of the gas in thermodynamic equilibrium. By the reciprocity theorem, we have

$$\left(\frac{\partial p}{\partial V}\right)_T \left(\frac{\partial V}{\partial T}\right)_p \left(\frac{\partial T}{\partial p}\right)_V = -1$$

or

$$\left(\frac{\partial p}{\partial T}\right)_V = -\left(\frac{\partial p}{\partial V}\right)_T \left(\frac{\partial V}{\partial T}\right)_p \qquad \text{...(i)}$$

The isothermal elasticity of the gas is defined by

$$E_T = V\left(\frac{\partial p}{\partial V}\right)_T$$

and the coefficient of volume expansion by

$$\alpha = \frac{1}{V}\left(\frac{\partial p}{\partial V}\right)_P$$

Hence eq. (i) becomes

$$\left(\frac{\partial p}{\partial T}\right)_V = E_T\alpha$$

The change in pressure dp corresponding to a change in temperature dT at constant volume is therefore given by

$$dp = E_T\alpha dT$$

$$= 10^5 \text{ dyne/cm}^2 \times 0.0037/\text{K} \times 0.1\text{K}$$

$$= 37 \text{ dyne/cm}^2.$$

2. *The pressure on 1 mole of mercury at 0°C is increased reversibly and isothermally from 0 to 1000 atmospheres. Calculate the heat transferred, the work done and the change in internal energy. Given : coefficient of volume expansion* $\alpha = 178 \times 10^{-6}$ *°C*$^{-1}$*, specific volume = 14.7 cm*^{3} */mole and isothermal compressibility* $k_T = 3.84 \times 10^{-12}$*cm*^{2}*/dyne. (1 atmosphere = 1.013 × 10*^{6} *dyne/cm*^{2}*).*

What would be the rise in temperature if the pressure were increased adiabatically ? The molar specific heat of mercury is 6.69 cal/mole-°C .

Solution : The heat transferred in isothermal compression is given by

$$\Delta Q = -\frac{TV\alpha}{J}(p_2 - p_1)$$

Here T= 0 + 273 = 273 K, V = 14.7 cm^3/mole,

α = 178 × 10^{-6}/°C = 178 x 10^{-6}/K and

$(p_2 - p_1)$ = (1000 – 0) = 1000 atmos

= 1000 × 1.013 x 10^6 dyne/cm^2

$$\therefore \Delta Q = \frac{(273\,\text{K})(14.7\,\text{cm}^3/\text{mole})(178\times10^{-6}/\text{K})1.013\times10^9\,\text{dyne/cm}^2}{4.18\times10^7\,\text{erg/cal}}$$

$$= -17.3 \text{ cal/mole} \qquad [\because \text{erg} = \text{dyne}\times\text{cm}]$$

The work done is

$$W = \int_{p_1}^{p_2} p\,dV = \int_{p_1}^{p_2} p\left(\frac{\partial V}{\partial p}\right)_T dp. \qquad \ldots\text{(i)}$$

Now, the isothermal compressibility (reciprocal of bulk modulus) is given by

$$K_T = -\frac{1}{V}\left(\frac{\partial V}{\partial P}\right)_T$$

so that

$$\left(\frac{\partial V}{\partial p}\right)_T = -Vk_T$$

Substituting this value in eq. (i), we get

$$W = -\int_{p_1}^{p_2} pVk_t\,dp$$

For liquids k_T is fairly independent of pressure.

$$\therefore \quad W = -Vk_T\int_{p_1}^{p_2} p\,dp = \frac{Vk_T}{2}\left(p_2^2 - p_1^2\right).$$

Substituting the given values :

$$W = -\frac{1}{2}\left(14.7\text{cm}^3/\text{mole}\right)\times\left(3.84\times10^{-12}\text{cm}^2/\text{dyne}\right)\times$$

$$\left(1000\times1.013\times10^6\text{dyne}/\text{cm}^2\right)^2$$

$$= -2.90\times10^7\text{erg}/\text{mole}$$

$$= \frac{2.90\times10^7}{4.18\times10^7} = -0.69\text{cal}/\text{mole}.$$

Thus, when one mole of mercury is compressed, 17.3 cal of heat is liberated but only 0.69 cal of work is done in compression. The extra amount of heat comes from the internal energy of mercury which is thus *decreased.*

∴ decrease in internal energy = 17.3 – 0.69 = 16.6 cal/ mole.

If the pressure were increased adiabatically, the heat liberated would increase the temperature of mercury. If s be the molar specific heat of mercury, then the rise in temperature is

$$\Delta T = \frac{\Delta Q}{s} = \frac{17.3\text{cal}/\text{mole}}{6.69\text{cal}/(\text{mole}-°\text{C})} = 2.58°\text{C}.$$

3. ***Calculate (a) the heat given out when the pressure on 200 gm of copper is increased reversibly and isothermally from zero to 1000 atm at 27°C, (b) the work done during the above compression, (c) the rise in temperature if the above compression were performed reversibly and adiabatically. For copper : density = 8.9 g/cm³, coefficient of volume expansion $\alpha = 49 \times 10^{-6}$/°C, isothermal compressibility $k_T = 0.776 \times 10^{-12}$ cm²/dyne and sp. heat = 0.093 cal/(g-°C). 1 atm = 1.01×10^6 dyne/cm² and J = 4.18×10^7 erg/cal.***

Solution : (a) The heat given out in isothermal compression is (in usual notations)

$$\Delta Q = -TV\alpha(p_2 - p_1)/J$$

$$= 300K \times \frac{200\,g}{8.9g/cm^3} \times \left(49 \times 10^{-6}/K\right) \times \left(1000 \times 1.01 \times 10^6 \frac{dyne}{cm^2}\right)$$

$= 4.18 \times 10^7$ erg/cal

$= -7.98$ cal.

(b) The work done is

$$W = -\frac{1}{2}Vk_T\left(p_2^2 - p_1^2\right)$$

$$= -\frac{1}{2}\left(\frac{200g}{8.9g/cm^3}\right)\left(0.776 \times 10^{-12}\frac{cm^2}{dyne}\right)\left(1000 \times 1.01 \times 10^6 \frac{dyne}{cm^2}\right)^2$$

$= -8.9 \times 10^6$ erg.

(c) If the compression were adiabatic, then the rise in temperature due to heat liberated is

$$\Delta T = \frac{\Delta Q}{mass \times sp.heat}$$

$$= \frac{7.98cal}{200gm \times 0.093cal/(g-{}^\circ C)} = 0.43^\circ C$$

4. Calculate the work done when pressure on 10 gram of copper is increased from 0 to 1000 atmospheres at 0°C. The density of copper is 8.93 gram/cm³ and the bulk modulus is 1.31 × 10¹² dyne/cm². (1 atmosphere = 101 × 10⁶ dyne/ cm²).

Solution : The work done on copper in compression from pressure p_1 *to* p_2, is given by

$$W = -\frac{1}{2}Vk_T\left(p_2^2 - p_1^2\right),$$

where V is volume and k_T is isothermal compressibility (reciprocal of bulk modulus) of copper. Substituting the given values (here $p_1 = 0$):

$$W = -\frac{1}{2}\left(\frac{10\text{gram}}{8.93\text{gram}/\text{cm}^3}\right)\left(\frac{1}{1.31\times10^{12}}\frac{\text{cm}^2}{\text{dyne}}\right)\left(1000\times1.01\times10^6\frac{\text{dyne}}{\text{cm}^2}\right)^2$$

$= - 0.0436 \times 10^7$ erg.

5. The pressure on 10 gram of water of 0 °C is increased reversibly and adiabatically from 0 to 1000 atmospheres. Calculate the change in temperature, given that the coefficient of volume expansion $\alpha = -67 \times 10^{-6}$ /°C, specific heat s = 1.0087 cal/(gram-°C) and specific volume = 1.000 cm^3/gram.

Solution : The heat produced in compression is

$$\Delta Q = TV\alpha(p_2 - p_1)/J.$$

Here $T = 0 + 273 = 273$ K,

$V = 1.000$ cm^3/gram × 10gm = 10.000 cm^3,

$\alpha = -67 \times 10^{-6}/°C = -67 \times 10^{-6}/K$,

$p_2 = 1000$ atmos $= 1000 \times 1.013 \times 10^6$ dyne/cm^2,

$p_1 = 0$ and $J = 4.18 \times 10^7$ erg/cal.

$$\therefore \quad \Delta Q = \frac{273\times10.000\times(-67\times10^{6})\times(1000\times1.013\times10^{6})}{4.18\times10^{7}}$$

$= - 4.43$ cal

The corresponding change in temperature is

$$\Delta T = \frac{\Delta Q}{mass \times sp.heat}$$

$$= \frac{-4.43\,cal}{10g\times1.0087\,cal/(g-°C)} = -0.44°C$$

The temperature drops by 0.44°C.

6. Calculate the molar heat capacity at constant volume Cv, the ratio C_p/C_v, and the adiabatic compressibility of mercury at 0°C and 1 atm pressure from the following data:

C_p = 6.69 cal/(mole-deg), V= 14.72 cm^3/mole, volume expansion coefficient α = 181 × 10^{-6}/°C, isothermal compressibility k_T = 3.88 × 10^{-12} cm^2/dyne.

Solution : The difference in molar heat capacities of any pure substance is given by

$$C_p - C_v = TE\alpha^2 V,$$

where T is Kelvin temperature, E the isothermal bulk modulus, α the coefficient of volume expansion and V the molar volume.

Now, $E = \frac{1}{k_T}$, where k_T is isothermal compressibility.

$$\therefore \quad C_p - C_y = \frac{T\alpha^2 V}{k_T}$$

Substituting the given values, we get

$$C_p - C_V = \frac{273\text{K} \times \left(181 \times 10^{-6}/\text{K}\right)^2 \times \left(14.72\text{cm}^3/\text{mole}\right)}{3.88 \times 10^{-12}\text{cm}^2/\text{dyne}}$$

$$= 3.39 \times 10^7 \text{dyne} - \text{cm}/(\text{mole} - \text{K})$$

$$= \frac{3.39 \times 10^7}{4.18 \times 10^7} = 0.81 \text{cal}/(\text{mole\_K})$$

$$\therefore \quad C_V = C_p - 0.81$$

$$= 6.69 - 0.81 = 5.88 \text{ cal/(mole-K)}.$$

$$\text{Now,} \quad \frac{C_p}{C_V} = \frac{6.69}{5.88} = 1.14$$

Finally, we know that

$$\frac{C_p}{C_V} = \frac{E_S}{E_T} = \frac{k_T}{k_S}$$

Where k_s, is adiabatic compressibility.

$$\therefore \quad k_S = \frac{k_T}{(C_p/C_V)} = \frac{3.88 \times 10^{-12}}{1.14} = 3.40 \times 10^{-12} \text{cm}^2/\text{dyne}.$$

7. At 0°C, aluminium has the following properties : atomic weight = 27.0 gram/mole; density = 2.70 gram/cm³, c_p = 0.220 cal/(gram-deg), α = 71.4 × 10^{-6}/deg, k_T = 1.34 × 10^{-12} cm²/dyne. Calculate, at 0°C (a) the molar heat capacity at constant volume, (b) the ratio γ, (c) the adiabatic compressibility.

Solution : *(a)* The difference in molar heat capacities of a pure substance is given by

$$C_p - C_V = TE\alpha^2 V = \frac{T\alpha^2 V}{k_T}$$

Now, molar volume $V = \frac{27.0 \text{ gram/mole}}{2.70 \text{ gram/cm}^3} = 10 \text{ cm}^3/\text{mole}$, and molar heat capacity at constant pressure C_p = 0.220cal/(gram-deg) × 27.0 gram/mole = 5.94 cal/(mole-deg).

$$\therefore \quad C_p - C_V = \frac{273 \text{deg} \times (71.4 \times 10^{-6}/\text{deg})^2 \times 10 \text{cm}^3/\text{mole}}{1.34 \times 10^{-12} \text{ cm}^2/\text{dyne}}$$

$$= 1.038 \times 10^7 \text{ dyne-cm/mole-deg}$$

$$= \frac{1.038 \times 10^7}{4.18 \times 10^7} \doteq 0.248 \text{ cal/mole-deg}$$

Thus
$$C_v = C_p - 0.248$$

$$= 5.94 - 0.248 = 5.69 \text{ cal/mole-deg.}$$

(b)
$$\gamma = \frac{C_p}{C_V} = \frac{5.94}{5.69} = 1.04$$

(c)
$$\frac{C_p}{C_V} = \frac{k_T}{k_S}$$

$$\therefore \quad k_s = \frac{k_T}{C_p/C_V} = \frac{1.34 \times 10^{-12}}{1.04} = 1.29 \times 10^{-12} \text{ cm}^2/\text{dyne.}$$

8. The tension in a copper wire of mass of 5 gm, diameter 1 mm and temperature 0°C is increased adiabatically from zero to 1000 newton. Calculate the change in temperature. For

copper : density = 8.93 gram/cm³, coefficient of linear expansion = 16.8 × 10^{-6} degree^{-1}, sp. heat = 0.091 cal gram^{-1} deg^{-1}.

Solution : The change in temperature in adiabatic stretching of a wire of mass per unit length w , coefficient of linear expansion α and specific heat s is given by

$$\Delta T = \frac{T\alpha}{Jws}\Delta F,$$

where ΔF is the stretching force and T is Kelvin temperature.

Here $T = 0 + 273 = 273$ K, $\alpha = 16.8 \times 10^{-6}$/K , $\Delta F = 1000$ newton = 10^8dyne, $w = \pi r^2 d = 3.14 \times (0.05 \text{ cm})^2 \times (8.93 \text{ g/cm}^3)$ = 0.070 g/cm, s = 0.091 cal/(g-K) and $J = 4.18 \times 10^7$erg/cal.

$$\therefore \quad \Delta T = \frac{273 \times (16.8 \times 10^{-6}) \times 10^8}{(4.18 \times 10^7) \times 0.070 \times 0.091}$$

$$= 1.7 \text{ K} = 1.7°\text{C}$$

Stretching means negative pressure. Hence the temperature will *fall* by 1.7°C .

9. A paramagnetic substance obeying Curie's law is placed in a magnetic field of 5000 gauss and a temperature of 2 K. What would be the temperature-change if the field is reduced reversibly and adiabatically to zero. Assume the Curie constant per gram to be 0.06 and the specific heat at constant field to remain constant and equal to 0.1 cal-gram^{-1}-deg^{-1}.

Solution : On adiabatic demagnetisation by withdrawing the magnetising field, the temperature-change is given by

$$\Delta T = \frac{(CV)}{2JC_H T}H^2,$$

where *(CV)* is the Curie's constant per gm, C_H is the specific heat, 7" the initial temperature and H the magnetising field. Substituting the given values, we have

$$\Delta T = -\frac{0.06}{2\times(4.18\times10^7)\times0.1\times2}\times(5.000)^2$$

$$= -0.1 \text{ K}$$

The negative sign indicates that the temperature will *fall.*

10. Calculate the cooling produced by the adiabatic demagnetisation of a paramagnetic substance as the field is reduced from 10,000 oersted to 0, the initial temperature being 2 K. Curie constant = 0.042 erg-deg/(gm-oersted²) and specific heat C_H = 0.42 joule/(gram-deg).

Solution: $\Delta T = \frac{(CV)}{2JC_H T}H^2,$

where C_H is sp. heat in cal/(gram-deg).

Here (CV) = 0.042 erg-deg/(gram-oersted²), H= 10,000 oersted, T =2 K, C_H = 0.42 joule/(gram-deg) = 0.42 × 10^7 erg/(gram-deg) = $\frac{0.42\times10^7}{J}$ cal/(gram-deg).

$$\therefore \quad \Delta T = \frac{0.042\times(10,000)^2}{2\times(0.42\times10^7)\times2}$$

$$= 0.25 \text{ K.}$$

9

Laws of Speed Distribution

Let us consider an assembly of N molecules of an ideal (non-interacting) gas contained in a volume V at absolute temperature T. These molecules are identical and sufficiently widely separated to be distinguished. Let n_1 molecules be of energy ε_1, n_2 molecules of energy ε_2 ...... and n_r molecules of energy ε_r. The most probable distribution of molecules among these r energies is subject to two conditions :

(i) The total number of molecules N is constant i.e.

$$N = n_1 + n_2 + ... + n_i + ... + n_r = \text{constant}$$

so that $$\delta N = \delta n_1 + \delta n_2 + ... + \delta n_i + ... + \delta n_r = 0$$

or $$\sum_i \delta n_i = 0 \qquad ...(i)$$

(ii) The total energy E of the assembly is constant *i.e.*

$$E = \varepsilon_1 n_1 + \varepsilon_2 n_2 + ... + \varepsilon_i n_i + ... + \varepsilon_r n_r = \text{constant}$$

or $$\delta E = \varepsilon_1 \delta n_1 + \varepsilon_2 \delta n_2 + \varepsilon_i \delta n_i + ... + \varepsilon_r \delta n_r = 0$$

or $$\sum_i \varepsilon_i \delta n_i = 0 \qquad ...(ii)$$

If the prior probability for a molecule to have the energy ε_i is g_i, then the probability for any distribution is

$$P = \frac{N!}{n_1!\ n_2!\ldots n_r!}(g_1)^{n_1}(g_2)^{n_2}\ldots.(g_r)^{n_r}$$

According to Maxwell-Boltzmann distribution law, the most probable distribution of molecules of an *i* deal gas in different energy states is given by

$$n_i = g_i e^{-\alpha} e^{-\beta\varepsilon_i}, \qquad \text{...(iii)}$$

where $e^{-\alpha}$ and β are constants.

Evaluation of the Constants $e^{-\alpha}$ ***and*** β **:** To do this, it is convenient to consider a continuous distribution of energies rather than discrete set $\varepsilon_1, \varepsilon_2, \ldots + \varepsilon_r$. The eq. (iii) then becomes

$$n(\varepsilon)\,d\varepsilon = g(\varepsilon)e^{-\alpha}e^{-\beta\varepsilon}\,d\varepsilon.$$

Here $n(\varepsilon)d\varepsilon$ represents the number of molecules having energies between ε and $\varepsilon + d\varepsilon$.

Since $\varepsilon = p^2/2m$, where p is the molecular momentum, the last equation may be written as

$$n(p)\,dp = g(p)e^{-\alpha}e^{-\beta p^2/2m}\,dp.$$

Here $n(p)\,dp$ represents the number of molecules having momenta between p and $p + dp$. The a priori probability $g(p)$ that a molecule have a momentum between p and $p + dp$ is equal to the number of cells in the phase space within which such a molecule may exist. If each cell has the infinitesimal volume h^3, then

$$g(p)\,dp = \frac{\iiiiiint dx\,dy\,dz\,dp_x\,dp_y\,dp_z}{h^3},$$

where the numerator represents the volume of the phase space occupied by the molecules with the specified momenta. Now,

$$\iiint dx\,dy\,dz = V,$$

where V is the volume occupied by the gas in ordinary position space.

Further, $\iiint dp_x\,dp_y\,dp_z$ is the volume between two spheres of radii p and $p + dp$ which contain all the molecules having momentum between p and $p + dp$ in the momentum space. Thus

$$\iiint dp_x\,dp_y\,dp_z = \frac{4}{3}\pi(p+dp)^3 - \frac{4}{3}\pi p^3$$

$$= 4\pi p^2 dp$$

$$\therefore g(p)dp = \frac{4\pi V p^2 dp}{h^3}$$

and
$$n(p)dp = \frac{4\pi V p^2 e^{-\alpha} e^{-\beta p^2/2m}}{h^3} dp. \qquad \text{...(iv)}$$

We can now evaluate $e^{-\alpha}$. Since $\int_0^\infty n(p)\,dp = N$ (total number of molecules), the integration of the last expression gives

$$N = \frac{4\pi V e^{-\alpha}}{h^3} \int_0^\infty p^2 e^{-\beta p^2/2m}\,dp$$

$$= \frac{V e^{-\alpha}}{h^3}\left(\frac{2\pi m}{\beta}\right)^{3/2}. \qquad \left[\because \int_0^\infty x^2 e^{-ax^2} dx = \frac{1}{4}\sqrt{\pi/a^3}\right]$$

$$\therefore\ e^{-\alpha} = \frac{N h^3}{V}\left(\frac{\beta}{2\pi m}\right)^{3/2}.$$

Substituting this value of $e^{-\alpha}$ in eq. (iv), we get

$$n(p)dp = 4\pi N\left(\frac{\beta}{2\pi m}\right)^{3/2} p^2 e^{-\beta p^2/2m} dp. \qquad \text{...(v)}$$

To find β, we compute the total energy E of the assembly of molecules.

Since $p^2 = 2m\varepsilon$ and $dp = \frac{m\,d\varepsilon}{\sqrt{2m\varepsilon}}$, eq. (v) can be written as

$$n(\varepsilon)d\varepsilon = 4\pi N\left(\frac{\beta}{2\pi m}\right)^{3/2} 2m\varepsilon\, e^{-\beta\varepsilon}\frac{m\,d\varepsilon}{\sqrt{2m\varepsilon}}$$

$$= \frac{2N\beta^{3/2}}{\pi^{1/2}}\varepsilon^{1/2} e^{-\beta\varepsilon} d\varepsilon \qquad \text{...(vi)}$$

The total energy E of the system is given by

$$E = \int_0^{r} \varepsilon n(\varepsilon)\,d\varepsilon$$

$$= \frac{2N\beta^{3/2}}{\pi^{1/2}}\int_0^{\infty}\varepsilon^{3/2}e^{-\beta\varepsilon}\,d\varepsilon$$

$$= \frac{3N}{2\beta}. \qquad \left[\because \int_0^{\infty} x^{3/2}e^{-ax}dx = \frac{3}{4a^2}\sqrt{\pi/a}\right]$$

From the kinetic theory of gas, the total energy E of N molecules of *an ideal* (non-interacting) gas at absolute temperature T is

$$E = \frac{3}{2}NkT,$$

where k is Boltzmann's constant. Comparing the last two equations, we get

$$\beta = \frac{1}{kT}$$

Substituting this value of β in eq. (vi), we get

$$\boxed{n(\varepsilon)d\varepsilon = 4\pi N\left(\frac{1}{\pi kT}\right)^{3/2}\varepsilon^{1/2}e^{-\varepsilon/kT}d\varepsilon} \qquad \text{...(vii)}$$

This equation gives the number of molecules with energies between ε and $\varepsilon + d\varepsilon$ in an ideal gas containing N molecules at absolute temperature T, and is called 'Maxwell distribution of energies'.

In order to find Maxwell's speed distribution law, we make the following substitution in eq. (vii):

$$\varepsilon = \frac{1}{2} mv^2 \quad \text{and} \quad d\varepsilon = mv\,dv.$$

This gives

$$\boxed{n(v)dv = 4\pi N\left(\frac{m}{2\pi kT}\right)^{3/2} v^2\, e^{-mv^2/2kT} dv.} \qquad \text{...(viii)}$$

This equation represents the number of molecules with speeds between v and $v + dv$ in an ideal gas having N molecules at absolute temperature T, and is called Maxwell's distribution of speeds.

***Maxwell's Velocity Distribution*:** This distribution is given by

$$n(\vec{v})d^3\vec{v} = N\left(\frac{m}{2\pi kT}\right)^{3/2} e^{-m\vec{v}^2/2kT} d^3\vec{v}$$

or

$$n(\vec{v})dv_x\, dv_y\, dv_z = N\left(\frac{m}{2\pi kT}\right)^{3/2} e^{-m(v_x^2+v_y^2+v_z^2)/2kT} dv_x\, dv_y\, dv_z.$$

$n(\vec{v})d^3\vec{v}$ is the number of molecules in the gas (containing N molecules at absolute temperature T) having velocities between $\vec{v}$ and $\vec{v} + d\vec{v}$, that is, x-components of velocities between v_x and v_x+dv_x, y-components between v_y and $v_y + dv_y$ and z-components between v_z and $v_z + dv_z$.

***Probability of a Gas Molecule having speed v at temperature T*:** The fraction of molecules having speeds between v and $v + dv$ is given by

$$\frac{n(v)dv}{N} = 4\pi\left(\frac{m}{2\pi kT}\right)^{3/2} v^2\, e^{-mv^2/2kT} dv.$$

By definition, this is the probability $P(v)\,dv$ that a molecule will have speed between v and $v + dv$. Thus

$$P(v)dv = 4\pi\left(\frac{m}{2\pi kT}\right)^{3/2} v^2\, e^{-mv^2/2kT}\,dv.$$

The probability that a molecule with have speed v, that is, the probability function $P(v)$ is given by

$$P(v) = 4\pi\left(\frac{m}{2\pi kT}\right)^{3/2} v^2\, e^{-mv^2/2kT}.$$

Maxwell Velocity Distribution

In an ideal non-atomic gas containing N molecules at absolute temperature T, the number of molecules per unit volume, which have a velocity between $\vec{v}$ and $\vec{v} + d\vec{v}$ (*i.e.* their x -components of velocity lie between v_x and v_x+dv_x, y - components between v_y and $v_y + dv_y$ and z -components between v_z and $v_z + dv_z$) is given by

$$\boxed{n(\vec{v})d^3\vec{v} = C\,e^{-mv^2/2kT}\,d^3\vec{v}.} \qquad \text{...(i)}$$

where $C = N'\left(\frac{m}{2\pi kT}\right)^{3/2}$, N' being the total number of molecules per unit volume, $d^3\vec{v} = dv_x\,dv_y\,dv_z$ and $\vec{v}^2 = v^2 = (v_x^2 + v_y^2 + v_z^2)$.

This relation is known as 'Maxwell's velocity distribution'.

Maxwell Distribution of a Velocity Component : From Maxwell velocity distribution, we can compute the number of molecules per unit volume, which have x -component of velocity in the range v_x to $v_x + dv_x$ (irrespective of the values of the other velocity components). This number is the sum of all the molecules having x component of velocity in this range, that is,

$$n'(v_x)dv_x = \int_{v_y}\int_{v_z} n(\vec{v})d^3\vec{v},$$

where the integration extend over all possible y and z velocity components of the molecules. Hence eq. (i) gives

$$\begin{aligned} n'(v_x)dv_x &= C\int_{v_y}\int_{v_z} e^{-m(v_x^2+v_y^2+v_z^2)/2kT} dv_x dv_y dv_z \\ &= Ce^{-mv_x^2/2kT} dv_x \int_{-\infty}^{+\infty}\int_{-\infty}^{+\infty} e^{-m(v_y^2+v_z^2)/2kT} dv_y dv_z \\ &= C'e^{-mv_x^2/2kT} dv_x, \end{aligned}$$

because the integration over all values of v_y and v_x gives merely some constant.

C′ can be evaluated by the normalizing condition that the summation of $n'(v_x)\, dv_x$ must yield N', the total number of molecules per unit volume.

$$\int_{-\infty}^{+\infty} n'(v_x)dv_x = C'\int_{-\infty}^{+\infty} e^{-mv_x^2/2kT} dv_x = N'$$

$$C' = N'\left(\frac{m}{2\pi kT}\right)^{1/2}$$

$$\therefore\ n'(v_x)dv_x = N'\left(\frac{m}{2\pi kT}\right)^{1/2} e^{-mv_x^2/2kT} dv_x \qquad \text{...(ii)}$$

The probability that a molecule has x-component of velocity between v_x and $v_x + dv_x$ is the ratio of such molecules to the total number, that is, $P(v_x) = n'(v_x)/N'$. *Thus*

$$P(v_x)dv_x = \left(\frac{m}{2\pi kT}\right)^{1/2} e^{-mv_x^2/2kT} dv_x \qquad \text{...(iii)}$$

This is the required expression.

(i) It is clear that the function $P(v_x)$ is distributed symmetrically about the value $v_x = 0$. The mean value of v_x for any molecule must always vanish :

$$\bar{v}_x = 0$$

This is physically clear since the x -component of velocity of a molecule is as likely to be positive as negative.

(ii) Again $P(v_x)$ has its maximum value when $v_x = 0$ and decreases rapidly as $|v_x|$ increases. Thus, the maximum value of $P(v_x)$ is $\left(\frac{m}{2\pi kT}\right)^{1/2}$

(iii) The average value of v_x is given by

$$\overline{v_x^2} = \frac{1}{N'}\int_{-\infty}^{+\infty} n'(v_x)\,dv_x \times v_x^2.$$

In view of eq. (ii), we have

$$\overline{v_x^2} = \left(\frac{m}{2\pi kT}\right)^{1/2}\int_{-\infty}^{+\infty} e^{-mv_x^2/2kT}\,v_x^2\,dv_x$$

Let $\frac{m}{2kT} = a$. then

$$\overline{v_x^2} = \left(\frac{a}{\pi}\right)^{1/2}\int_{-\infty}^{+\infty} e^{-av_x^2}\,v_x^2\,dv_x$$

The value of the definite integral is $\frac{1}{2a}\sqrt{\frac{\pi}{a}}$.

$$\therefore \quad \overline{v_x^2} = \left(\frac{a}{\pi}\right)^{1/2}\frac{1}{2a}\left(\frac{\pi}{a}\right)^{1/2}$$

$$= \frac{1}{2a} = \frac{kT}{m}.$$

Using Maxwell's velocity distribution law, find v_x for which the probability falls to 1/e times its maximum value. *Also deduce expressions for $\overline{v_x}$ and $\overline{v_x^2}$.* We have shown above that the probability of a molecule of having x -component of velocity between v_x and v_x+dv_x is given by

$$P(v_x)dv_x = \left(\frac{m}{2\pi kT}\right)^{1/2} e^{-mv_x^2/2kT}\,dv_x$$

Therefore, the probability that a molecule has the x -component of velocity v_x, that is, the probability function $P(v_x)$ is

$$P(v_x) = \left(\frac{m}{2\pi kT}\right)^{1/2} e^{-mv_x^2/2kT}$$

The maximum value of the exponential term is 1. Therefore

$$\{P(v_x)\}_{\max} = \left(\frac{m}{2\pi kT}\right)^{1/2}$$

Thus $$\frac{P(v_x)}{\{P(v_x)\}_{\max}} = e^{-mv_x^2/2kT}$$

Given that $P(v_x)/\{P(v_x)_{\max} = \frac{1}{e} = e^{-1}$

$$\therefore \quad mv_x^2/2kT = 1$$

or $$v_x = \sqrt{2kT/m}$$

The expressions for $\overline{v_x}$ and $\overline{v_x^2}$ have been obtained in the last question ($\overline{v_x} = 0$ and $\overline{v_x^2} = kT/m$)

Classical and Quantum Statistics

Every solid, liquid or gas is an assembly of an enormous number of microscopic particles Likewise, radiation is an assembly of photons. Obviously, the actual motions or interactions of individual particles cannot be investigated. However, the macroscopic properties of such assemblies can be explained in terms of the statistical distribution of the individuals among different possible states and their most probable behaviour.

For example, from Maxwell distribution of speeds among the molecules of a gas we can calculate mean speed (which is related to the momentum carried by the molecules), mean-square speed (which is related to the energy of the molecules), and so on. From these *average* quantities we calculate observable properties like pressure and temperature of the gas.

Usually we consider how a fixed amount of energy is distributed among the various *identical* particles of an assembly. Now, there are three kinds of *identical* particles:

(i) Identical particles of any spin which are so much separated in the assembly that they *can be distinguished* from one another. The molecules of a gas are particles of this kind.

(ii) Identical particles of zero or integral spin which *cannot be distinguished* from one another. These are called Bose particles (or bosons) and do not obey Pauli's exclusion principle. Photons, phonons and α-particles are of this kind.

(iii) Identical particles of half-integral spin which *cannot be distinguished* from one another. These are called Fermi particles (or fermions) and do obey Pauli's exclusion principle. Electrons, protons, neutrons are particles of this kind.

The first kind of the particles are the classical particles and obey the Maxwell-Boltzmann energy distribution law. The second and third kinds of the particles are quantum particles and energy distribution law for them can be derived by methods of quantum statistics only.

Maxwell-Boltzmann (Classical) Statistics

Let us consider an assembly of N molecules (distinguishable identical particles). Suppose the energies of the molecules are limited to the values $\varepsilon_1, \varepsilon_2, ..., \varepsilon_r$ which represent either discrete energy states or average energies within a sequence of energy intervals. Let us divide the whole volume of the phase space into r cells and distribute the N molecules among these cells. Let us consider a distribution such that n_1 molecules (each having energy ε_1) are in cell 1, n_2 molecules (each having energy ε_2) are in cell 2, and so on. The probability P for the distribution is given by

$$P = \frac{N!}{n_1!\, n_2! \ldots n_r!} (g_1)^{n_1} (g_2)^{n_2} \ldots\ldots\ldots (g_r)^{n_r}$$

$$= \frac{N!}{\prod_{i=1}^{r} (n_i!)} \prod_{i=1}^{r} (g_i)^{n_i} \qquad [\prod \text{ stands for product}]$$

$$= N! \prod_{i=1}^{r} \left[\frac{(g_i)^{n_i}}{n_i!} \right].$$

where n_i is the number of molecules having energy ε_i and g_i is the number of energy levels each having energy ε_i.

There are two constraints. The total number of molecules is fixed at N. That is,

$$\sum_{i=1}^{r} n_i = n_1 + n_2 + \ldots + n_r = N.$$

At a constant temperature, the total energy is fixed at E (say). That is,

$$\sum_{i=1}^{r} \varepsilon_i n_i = \varepsilon_1 n_1 + \varepsilon_2 n_2 + \ldots + \varepsilon_r n_r = E.$$

The next step is to determine the most probable distribution of the molecules which would correspond to the equilibrium state. This distribution would yield maximum value of P. Thus, for the equilibrium state, we have

$$\delta P = 0$$

subject to the limitations

$$\Sigma \delta n_i = 0$$

and

$$\Sigma \varepsilon_i \delta n_i = 0.$$

This leads to the result

$$\boxed{n_i = \frac{g_i}{e^{\alpha} e^{\varepsilon_i / kT}},}$$

where e^{α} is a constant, k is Boltzmann's constant and T is absolute temperature.

This result, known as Maxwell-Boltzmann distribution law, explains the observed properties of a gas.

The M-B distribution law, however, failed to explain the energy distribution in electromagnetic spectrum and also the properties (such as specific heat, conductivity) of metals. In other words, it cannot explain the behaviour of photons or of electrons in a metal. The basic reason is that gas molecules obeying M-B distribution can be distinguished from one another in some way and as such they can be given names. On the other hand, photons or electrons cannot be distinguished from one another and as such there is no way to give them names. The interchange of two photons, or of two electrons, does not result in any new configuration. Hence they need distribution law other than M-B.

Bose-Einstein Quantum Statistics

This statistics applies to identical, indistinguishable particles which do not obey Pauli's exclusion principle (for example, photons).

The basic assumption in B-E statistics is that any number of particles can be in any quantum state and that all quantum states are equally probable. Suppose each quantum state corresponds to an elementary cell in the phase space. The number of different ways in which n_i *indistinguishable* particles can be distributed among g_i cells is

$$\frac{(n_i + g_i - 1)!}{n_i!(g_i - 1)!}.$$

When the number of cells is sufficiently large, we can write the above expression as

$$\frac{(n_i + g_i)!}{n_i!\, g_i!}$$

The probability P of the entire distribution of N particles is the *product* of the number of different arrangements of particles among the states having each energy. Thus

$$P = \prod_i \frac{(n_i + g_i)!}{n_i!\, g_i!}$$

For the equilibrium (most probable) state, we must have

$$\delta P = 0$$

subject to the limitations

$$\sum \delta n_i = 0$$

and

$$\sum \varepsilon_i \delta n_i = 0 .$$

This leads to the result

$$\boxed{n_i = \frac{g_i}{e^{\alpha} e^{\varepsilon_i / kT} - 1}}.$$

This is Bose-Einstein distribution law. It explains successfully the laws of black-body radiation.

Fermi-Dirac Quantum Statistics

This statistics applies to identical, indistinguishable particles which do aboy Pauli's exclusion principle (for example, electrons).

The general method of deriving the Fermi-Dirac distribution law is similar to that for the Bose-Einstein distribution law except that now each phase cell (i.e. each quantum state) can be occupied by at most one particle. On this basis, we obtain

$$\boxed{n_i = \frac{g_i}{e^{\alpha} e^{\varepsilon_i / kT} + 1}}.$$

Again, n_i is the number of particles having energy ε_i, and g_i is the number of states having same energy ε_i.

Comparison of the three Distribution Laws : Let us define a quantity

$$f(\varepsilon_i) = \frac{n_i}{g_i}.$$

$f(\varepsilon_i)$ is called the 'occupation index' of a state of energy ε_i. It represents the average number of particles in each of the states of that energy. Thus

$$f_{MB}(\varepsilon_i) = \frac{1}{e^{\alpha} e^{\varepsilon_i/kT}},$$

$$f_{BE}(\varepsilon_i) = \frac{1}{e^{\alpha} e^{\varepsilon_i/kT} - 1}$$

and

$$f_{FD}(\varepsilon_i) = \frac{1}{e^{\alpha} e^{\varepsilon_i/kT} + 1}$$

In the following diagrams we plot each occupation index $f(\varepsilon_i)$ against energy ε_i for two different values of T and α.

In the M-B distribution the occupation index $f(\varepsilon_i)$ falls purely *exponentially* with increase in energy ε_i. In fact, it falls by a factor of $1/e$ for each increase of kT in the energy ε_i.

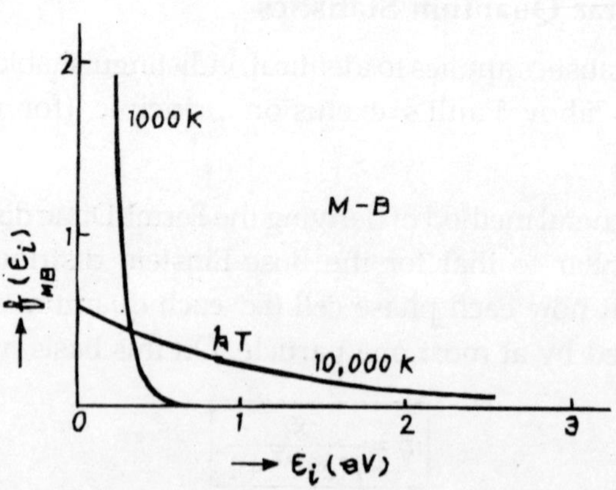

Fig.

The B-E occupation index against energy is plotted in Fig. for temperatures of 1000 K and 10,000 K, in each case for

$\alpha = 0$ (corresponding to a "gas" of photons). For $\varepsilon_i >> kT$, the B-E distribution approaches the exponential form characteristic of the M-B distribution. In this region the average number of

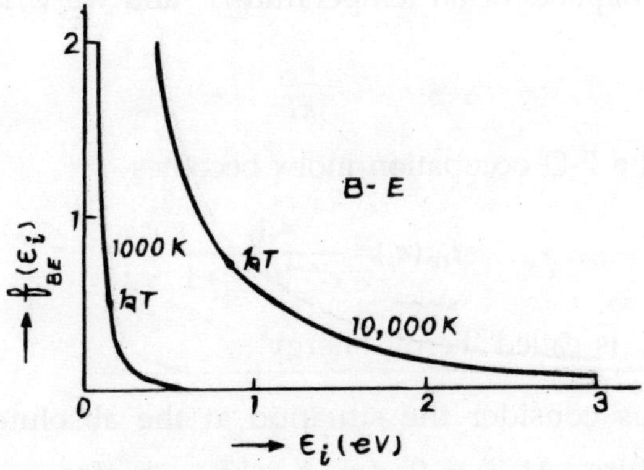

Fig.

particles per quantum state is much less than 1. However, for $\varepsilon_i << kT$, *the* -1 term in the denominator causes the occupation index much greater compared to that of the M-B distribution. This means that for energies small compared to kT, the number of particles per quantum state is greater for the B-E distribution than for the M-B distribution.

The F-D distribution is plotted in Fig. for three different values of T and α. In this distribution the occupation index never goes above 1. This

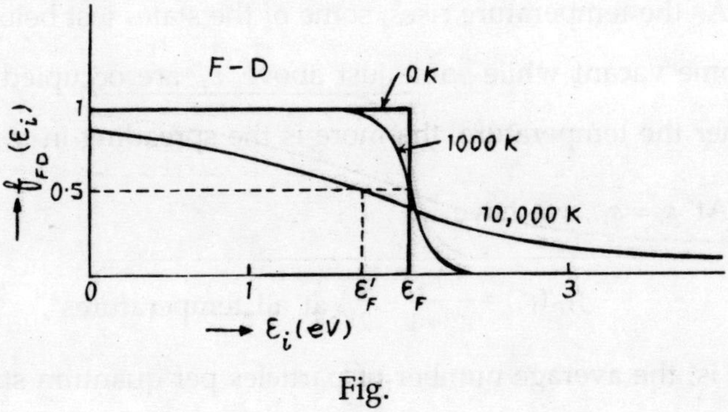

Fig.

signifies that we cannot have more than 1 particle per quantum state as required by Pauli's exclusion principle which applies in this case. Further, in this distribution, the parameter α is strongly dependent on temperature *T*, and we write

$$\alpha = \frac{\varepsilon_F}{kT}$$

so that the F-D occupation index becomes

$$f_{FD}(\varepsilon_i) = \frac{1}{e^{(\varepsilon_i - \varepsilon_F)/kT} + 1}$$

where ε_F is called 'Fermi Energy'.

Let us consider the situation at the absolute zero of temperature. At $T = 0$, $(\varepsilon_i - \varepsilon_F)/kT = -\infty$ (for $\varepsilon_i < \varepsilon_F$) and $(\varepsilon_i - \varepsilon_F)/kT = +\infty$ (for $\varepsilon_i > \varepsilon_F$). Therefore

for $\varepsilon_i < \varepsilon_F$ $\qquad f_{FD}(\varepsilon_i) = \dfrac{1}{e^{-\infty} + 1} = 1 \qquad (\because e^{-\infty} = 0)$

and for $n_i = \dfrac{g_i}{e^{\alpha} e^{\varepsilon_i/kT} - 1}$ $\qquad f_{FD}(\varepsilon_i) = \dfrac{1}{e^{+\infty} + 1} = 1 \qquad (\because e^{+\infty} = \infty)$

Thus, at $T = 0$, all energy states from $\varepsilon_i = 0$ to $\varepsilon_i = \varepsilon_F$ are occupied because $f_{FD}(\varepsilon_i) = 1$. while all states above ε_F are vacant.

As the temperature rises, some of the states just below ε_F become vacant while some just above ε_F are occupied. The higher the temperature, the more is the spreading in $f_{FD}(\varepsilon_i)$.

At $\varepsilon_i = \varepsilon_F$, we have

$$f_{FD}(\varepsilon_i) = \frac{1}{e^0 + 1} = \frac{1}{2}, \text{ at all temperatures*},$$

that is, the average number of particles per quantum state is exactly $\frac{1}{2}$. In other words, the probability of finding an electron with energy equal to the Fermi energy in a metal is $\frac{1}{2}$ at any temperature.

Comparison of the three Statistics

Maxwell-Boltzmann (classical)	Bose-Einstein (quantum)	Fermi-Dirac (quantum)
1. Applies to systems of distinguishable particles, such as gases.	Applies to systems of indistinguishable particles not obeying Pauli's exclusion principle, such as photons, phonos and liquid Helium.	Applies to indistinguishable particles obeying Pauli's exclusion principle, such as electrons.
2. There is no restriction on the number of particles in a given state.	There is no restriction on the number of particles in a given state.	Only one particle in a given quantum state is allowed.
3. In the phase space the volume of a phase cell is not fixed.	The volume of phase cell is of the order of h^3.	The volume of phase cell is of the order of h^3.
4. The distribution law is $n_i = \frac{g_i}{e^{\alpha} e^{\varepsilon_i/kT}}$	The distribution law is $n_i = \frac{g_i}{e^{\alpha} e^{\varepsilon_i/kT} + 1}$	The distribution law is $n_i = \frac{g_i}{e^{\alpha} e^{\varepsilon_i/kT} + 1}$
5. The behavior of distribution function $f(\varepsilon_i)$ against ε_i is exponential.	For $\varepsilon_i >> kT$, exponential; For $\varepsilon_i << kT$, lies above M-B.	For $\varepsilon_i >> kT$, exponential where $\varepsilon_i >> \varepsilon_F$. If $\varepsilon_F >> kT$, decreases Abruptly near ε_F
6. The energy of an M-B system, such as in ideal monatomic gas, may be zero at absolute zero.	The energy may be zero at absolute zero.	The energy at absolute zero cannot be zero because all particles cannot come down to the ground state due to Pauli's exclusion principle.

Limits when both B-E and F-D approach M-B Statistics:

The distribution law of the three statistics are as under :

$$\frac{g_i}{n_i} = e^{\alpha} e^{\varepsilon_i/kT} \qquad \text{(M-B)}$$

$$\frac{g_i}{n_i} = e^{\alpha} e^{\varepsilon_i/kT} - 1 \qquad \text{(B-E)}$$

$$\frac{g_i}{n_i} = e^{\alpha} e^{\varepsilon_i/kT} + 1 \qquad \text{(F-D)}$$

If $\frac{g_i}{n_i} >> 1$, then $\frac{g_i}{n_i} \approx \left(\frac{g_i}{n_i}+1\right) \approx \left(\frac{g_i}{n_i}-1\right)$. In this limit both the B-E and the F-D distributions are identical with the M-B distribution. This limit ($g_i / n_i >> 1$) occurs when the temperature is not too low and the pressure (or density) is not too high.

PROBLEMS

1. There are a large number of particles each of mass 0.1 g, all lying in a box at an equilibrium temperature 300K. Calculate the probability that any one of them will spontaneously fly to a height 1 Å. Assume that they obey M.B. statistics (k = 1.38 x 10^{-23} joule/K).

Solution : In M-B distribution, the probability of a particle for having an energy e is given by

$$P(\varepsilon) = 2\pi\left(\frac{1}{\pi kT}\right)^{3/2} \varepsilon^{1/2} e^{-\varepsilon/kT}$$

Here, the particle, in order to fly to a height 1 Å, must have energy

$$\varepsilon = mgh = (0.1 \times 10^{-3}\text{ kg})\ (9.8\text{ N/kg})\ (10^{-10}\text{ m})$$

$$= 9.8\text{x}10^{-14}\text{ joule.}$$

Also, $kT = (1.38 \times 10^{-23}\text{ joule/K})\ (300\text{ K})$

$$= 4.14\text{x } 10^{-21}\text{ joule.}$$

$$\therefore \frac{\varepsilon}{kT} = \frac{9.8 \times 10^{-14}\text{ joule}}{4.14 \times 10^{-21}\text{ joule}} = 2.37 \times 10^{7}$$

and $e^{\varepsilon/kT} = \text{e}^{-2.37 \times 10^{7}} \approx 0$

The probability is almost **zero**.

2. *Calculate the probability that the speed of an oxygen molecule lies between 100 and 101 meter/sec at a temperature of 200 K. The mass of an oxygen molecule is 5.3 x 10^{-26} kg. (k = 1.38 × 10^{-23} joule/K.)*

Solution : Under M.B. distribution, the probability that a molecule has speed between v and $v + dv$ is given by

$$P(v)dv = 4\pi\left(\frac{m}{2\pi kT}\right)^{3/2} v^2 e^{-mv^2/kT} dv.$$

Here m = 5.3x 10^{-26} kg,

v = 100 m/s,

dv = 101-100= 1 m/s

and T = 200K.

$$\therefore \frac{m}{2\pi kT} = \frac{5.3 \times 10^{-26}}{2 \times 3.14 \times (1.38 \times 10^{-23}) \times 200} = 3 \times 10^{-6}$$

and $$\frac{mv^2}{2kT} = \frac{(5.3 \times 10^{-26}) \times (100)^2}{2 \times (1.38 \times 10^{-23}) \times 200} = 0.1$$

$$\therefore P(v)\, dv = 4 \times 3.14 \times (3 \times 10^{-6})^{3/2} \times (100)^2 \times e^{-0.1} \times 1$$
$$= 4 \times 3.14 \times (5.2 \times 10^{-9}) \times 10^4 \times 0.90 \simeq 6 \times 10^{-4}.$$

3. *An assembly has only two particles and there are only two phase cells, or quantum states, to be occupied. Show the various possible arrangements in (i) Maxwell-Boltzmann, (ii) Bose-Einstein and (iii) Fermi-Dirac statistics.*

Solution : (i) In M-B (classical) statistics, the particles are distinguishable and can be named, say a and b ; and there is no restriction on the number of particles in a cell. This means that each or both of them can occupy any one of the two cells. Thus, we have the following four possible arrangements :

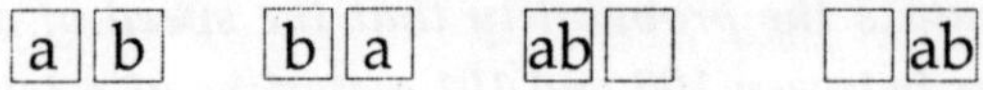

(M-B DISTRIBUTION)

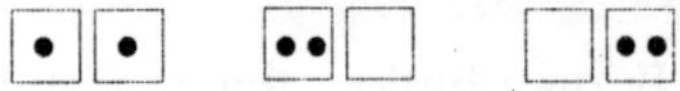

(B -E DISTRIBUTION)

(F - D DISTRIBUTION)

(Fig.)

(ii) In B-E (quantum) statistics, the particles are indistinguishable and hence cannot be named. Therefore, we must consider the cells to be occupied, rather than the particles. However, there is no restriction on the number of particles in a cell. Thus, in this case, we have the above three arrangements.

(iii) In F-D (quantum) statistics, again the particles are indistinguishable and, in this case, only one particle can occupy one cell. Hence there is only one possible arrangement.

4. An assembly has only two particles which are to be placed in three phase cells. Compute the possible number of arrangements in (i) M-B, (ii) B-E and (iii) F-D statistics.

(i) nine, (ii) six, (iii) three.

10

Physical Application

Fourier Series Analysis

(a) *Analysis of a Square Wave by Foureirs Theorem*

Let us consider any physical quantity y (such as displacement, pressure or electric current) varying periodically as shown in fig. Here y has a constant value *a* from time *t=0* to *t=T/2* and then remains zero for the remainder of its period T. We may express this function as follows :

$$\left.\begin{aligned} & y = a \text{ from } t = 0t = T/2 \\ \text{and } & y = 0 \text{ from } t = 0t = T/2 \end{aligned}\right\} \quad ...(i)$$

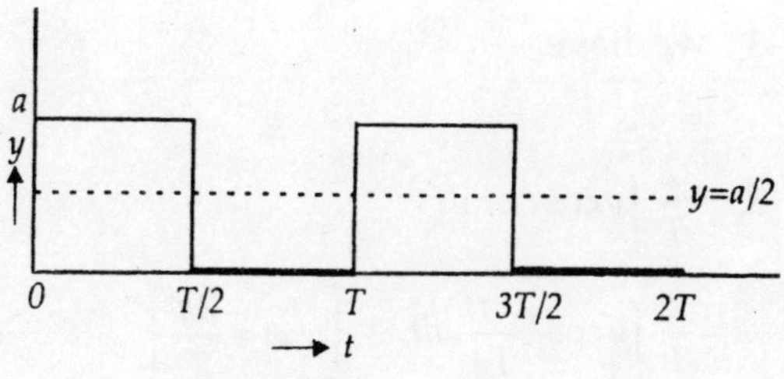

Fig.

when the time-axis of coordinates is taken through the lowest point of the displacement curve.

The Fourier's theorem can be expressed as

$$y = f\ (t) = A_0 + A_1 \cos \omega t + \ldots + A_r \cos r\omega t + \ldots$$

$$+B_1 \sin \omega t + \ldots + B_r \sin r\omega_t + \ldots \qquad \text{...(ii)}$$

where $A_0 = \frac{1}{T}\int_0^T y dt, A_r = \frac{2}{T}\int_0^T y \cos r\omega t\ dt$

and $B_r = \frac{2}{T}\int_0^T y \sin r\omega t\ dt$

Let us obtain the values of A_0, A_r and B_r for the given function:

$$A_0 = \frac{1}{T}\int_0^T y dt.$$

Substituting the value of y from eq. (i) we get

$$A_0 = \frac{1}{T}\int_0^{T/2} a\ dt.$$

$$= \frac{a}{T}[t]_0^{T/2} = \frac{a}{2},$$

i.e., the axis of the displacement curve is the line $y = \frac{a}{2}$ (shown dotted in fig.).

For A_r, we have

$$A_r = \frac{2}{T}\int_0^T y \cos r\omega t\ dt$$

$$= \frac{2}{T}\int_0^T y \cos \frac{2r\pi t}{T}\ dt. \qquad \left[\because \omega = \frac{2\pi}{T}\right]$$

Substituting the value of y from eq. (i), we have

$$A_r = \frac{2}{T}\int_0^{T/2} a\cos\frac{2r\pi t}{T}\,dt$$

$$= \frac{2a}{T}\frac{T}{2r\pi}\left[\sin\frac{2r\pi t}{T}\right]_0^{T/2}$$

$$= \frac{a}{r\pi}[\sin r\pi - \sin 0]$$

$$= 0.$$

Hence all the cosine terms in eq. (ii) are zero.

Similarly, for b_r, we have

$$B_r = \frac{2}{T}\int_0^{T/2} y\sin r\omega t\,dt$$

$$= \frac{2}{T}\int_0^{T/2} y\sin\frac{2r\pi t}{T}\,dt.$$

Substituting the value of y from eq. (i) we have

$$B_r = \frac{2}{T}\int_0^{T/2} a\sin\frac{2r\pi t}{T}\,dt$$

$$= \frac{2a}{T}\frac{T}{2r\pi}\left[-\cos\frac{2r\pi t}{T}dt\right]_0^{T/2}$$

$$= \frac{a}{r\pi}[-\cos r\pi - \cos 0]$$

$$= \frac{a}{r\pi}[-\cos r\pi + 1]$$

when r is even, $\cos r\pi = +1$ and when r is odd, $\cos r\pi = -1$. Thus

$$B_r = \frac{a}{r\pi}\left[-\begin{Bmatrix}1 \text{ for even } r\\ -1 \text{ for odd } r\end{Bmatrix}+1\right]$$

$$= \frac{a}{r\pi}\left[(-1)^r + 1\right] = \frac{a}{r\pi}\left[1 + (-1)^{r+1}\right].$$

Putting $r = 1, 2, 3, \ldots$, we get

$$B1 = \frac{2a}{\pi} B_2 = 0, B_3 = \frac{2a}{3\pi}, B_4 = 0, B_5 = \frac{2a}{5\pi}, \ldots$$

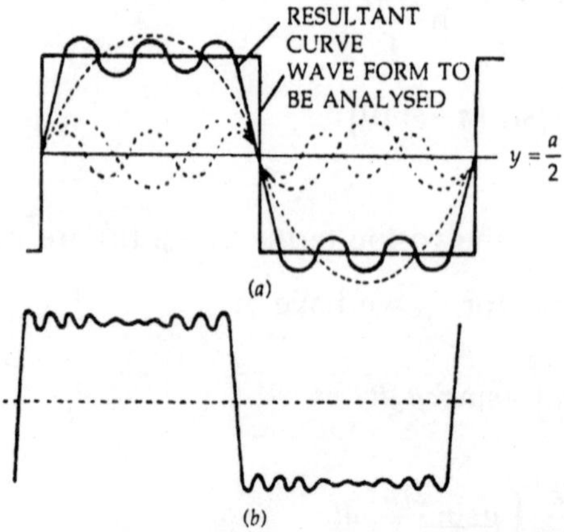

Fig.

Therefore, putting the values of A_0, A_r and B_r, eq. (ii) becomes

$$y = \frac{a}{2} + \frac{2a}{\pi}\left[\sin\omega t + \frac{1}{3}\sin 3\omega t + \frac{1}{5}\sin 5\omega t + \ldots\right]$$

$$or\ \ y = \frac{a}{2} + \frac{2a}{\pi}\left[\sin\frac{2\pi t}{T} + \frac{1}{3}\sin\frac{6\pi t}{T} + \frac{1}{5}\sin\frac{10\pi t}{T} + \ldots\right]$$

Thus the given complex motion has the axis $y = a/2$ and the various components as under :

$$\frac{2a}{\pi}\sin\frac{2\pi t}{T}$$

$$\frac{2a}{3\pi}\sin\frac{6\pi t}{T}$$

$$\frac{2a}{5\pi}\sin\frac{10\pi t}{T}, \text{and so on,}$$

having frequencies in the ratio 1: 3: 5 etc. and amplitudes in the ratio

$$1:\frac{1}{3}:\frac{1}{5} \text{ etc.}$$

In Fig., the addition of the successive terms is indicated graphically. In fig. (a) the first three terms of the series are shown independently and also the resultant curve. It is seen that even in this case the resultant curve roughly resembles the curve to be analysed. By the addition of 15 terms of the series as shown in fig. (b); the curve is almost perfectly reproduced.

Analysis of a Harmonic Function (Square Wave) by Fourier's Theorem

The given harmonic function

$$\left.\begin{aligned} & y = a \text{ from } t = 0 \text{ to } t = T/2 \\ \text{and } & y = -a \text{ from } t = T/2 \text{ to } t = T \end{aligned}\right\} \quad \ldots\text{(i)}$$

represents a square wave as shown in fig.

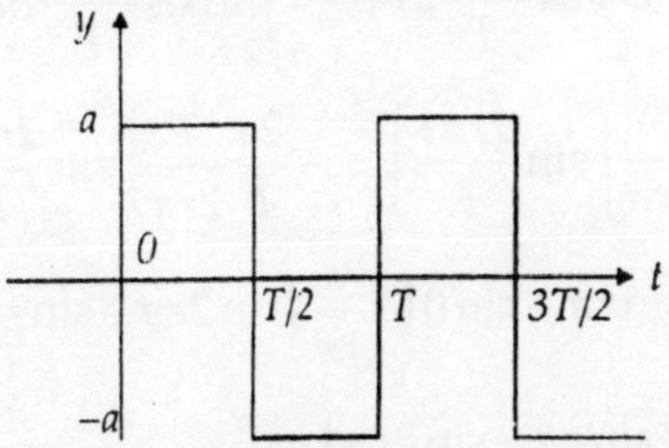

Fig.

The Fourier series of a harmonic function is

$$\begin{aligned} y &= f(t) = A_0 + A_1 \cos \omega t + \ldots + A_r \cos r\omega\, t + \ldots \\ &= B_1 \sin \omega t + \ldots + B_r \sin r\omega t + \ldots, \end{aligned}$$

where $A_0 = \frac{1}{T}\int_0^T y\,dt, A_r = \frac{2}{T}\int_0^T y \cos r\omega\, t\, dt$

and $B_r = \frac{2}{T}\int_0^T y \sin r\omega\, t\, dt.$

Let us obtain the values of A_0, A_r and B_r for the given function:

$$A_0 = \frac{1}{T}\int_0^T y\,dt$$

Substituting the values of y from eq. (i), we get

$$A_0 = \frac{1}{T}\int_0^T y\,dt - \frac{1}{T}\int_{T/2}^T a\,dt$$

$$= \frac{a}{T}[t]_{T/2}^T$$

Thus the axis of the displacement curve is the x-axis ($y=0$). For A_r, we have

$$A_r = \frac{2}{T}\int_0^T y\,\cos(r\omega t)t$$

$$= \frac{2}{T}\int_0^T y\cos\frac{2r\pi t}{T}dt. \qquad \left[\because w = \frac{2\pi}{T}\right]$$

Substituting the value of y from eq. (i), we get

$$Ar = \frac{2}{T}\int_0^{T/2} a\,\cos\frac{2r\pi t}{T}dt - \frac{2}{T}\int_{T/2}^T a\cos\frac{2r\pi t}{T}dt$$

$$= \frac{2a}{T}\frac{T}{2r\pi}\left[\sin\frac{2r\pi t}{T}\right]_0^{T/2} - \frac{2a}{T}\frac{T}{2r\pi}\left[\sin\frac{2r\pi t}{T}\right]_{T/2}^T$$

$$= \frac{a}{r\pi}[\sin r\pi - \sin 0] - \frac{a}{r\pi}[\sin 2r\pi - \sin r\pi]$$

$$= \frac{a}{r\pi}[2\sin r\pi - \sin 2r\pi] = 0.$$

Hence all the cosine term in eq. (ii) are zero. For B_r, we have

$$B_r = \frac{2}{T}\int_0^t y\sin(r\omega t)\,dt$$

$$= \frac{2}{T}\int_0^T y\sin\frac{2r\pi t}{T}dt$$

$$= \frac{2}{T}\int_0^{T/2} a\sin\frac{2r\pi t}{T}dt - \frac{2}{T}\int_{T/2}^T a\sin\frac{2r\pi t}{T}dt$$

$$= \frac{2a}{T}\frac{T}{2r\pi}\left[-\cos\frac{2r\pi t}{T}\right]_0^{T/2} - \frac{2a}{T}\frac{T}{2r\pi}\left[-\cos\frac{2r\pi t}{T}\right]_{T/2}^{T}$$

$$= \frac{a}{r\pi}[-\cos r\pi + \cos 0] - \frac{a}{r\pi}[-\cos 2r\pi + \cos r\pi]$$

$$= \frac{a}{r\pi}[-\cos r\pi + 1 + \cos 2r\pi - \cos r\pi]$$

$$= \frac{a}{r\pi}[2 - 2\cos r\pi] \qquad [\because \cos 2r\pi = 1]$$

$$= \frac{2a}{r\pi}[1 - \cos r\pi].$$

But $\cos\ r\pi = (-1)^r$, where $r = 1, 2, 3, \ldots$

$$\therefore B_r = \frac{2a}{r\pi}\left[1-(-1)^r\right].$$

Putting r = 1, 2, 3,...we get

$$B_1 = \frac{4a}{\pi}, B_2 = 0, B_3 = \frac{4a}{3\pi}, B_4 = 0 B_5 = \frac{4a}{5a}, \ldots$$

Putting the values of A_0, $A_{r'}$ and $B_{r'}$ in eq. (ii), we get

$$y = \frac{4a}{\pi}\left[\sin\omega t + \frac{1}{3}\sin 3\omega t + \frac{1}{5}\sin 5\omega t + \ldots\right]$$

or $$y = \frac{4a}{\pi}\left[\sin\frac{2\pi t}{T} + \frac{1}{3}\sin\frac{2\pi t}{T} + \frac{1}{5}\sin\frac{10\pi t}{T} + \ldots\right]$$

Thus the given function can be written as a sum of simple harmonic components having frequencies in the ratio 1 : 3 : 5 : .....and amplitudes in the ratio $1 : \frac{1}{3} : \frac{1}{5} : \ldots$

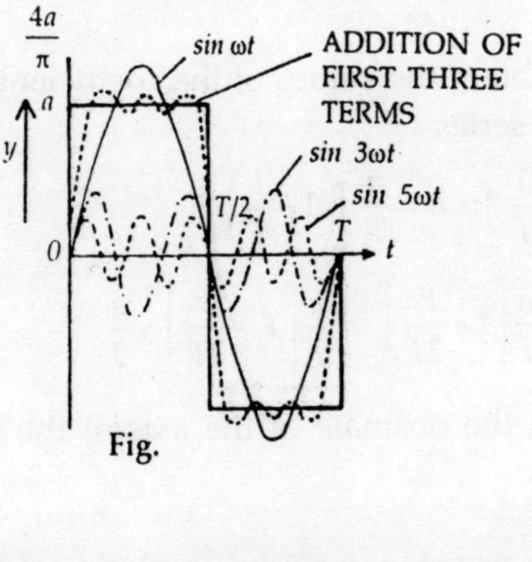

Fig.

If we take the first three terms of the above series representing the square wave and add them together, the result is Fig. The First harmonic has the frequency of the square wave and the higher frequencies build up the squareness of the wave. The highest frequencies are responsible for the sharpness of the vertical sides of the wave.

(b) *Analysis of a Saw-Tooth Wave by Fourier's Theorem*

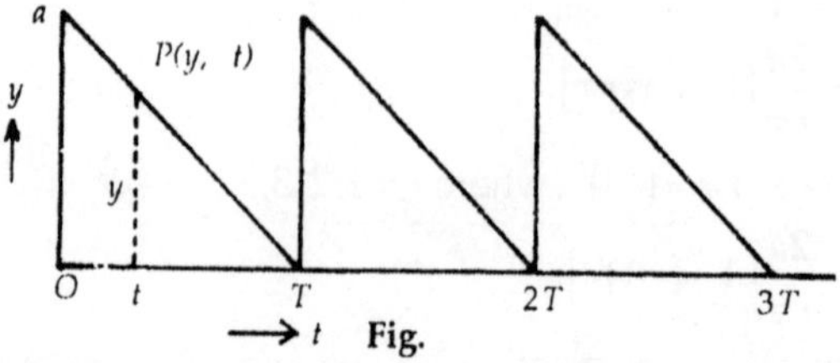

Fig.

The curve shown in fig. represents a saw-tooth wave. It can be expressed mathematically as

$$y = f(t) = a\left(1 - \frac{t}{T}\right), \text{ from } t = 0 \text{ to } t = T. \qquad \ldots\text{(i)}$$

The Fourier's series is

$$y = f(t) = A_0 + A_1 \cos\omega t + \ldots + A_r \cos r\omega t + \ldots$$
$$+ B_1 \sin\omega t + \ldots + B_r \sin r\omega t + \ldots. \qquad \ldots\text{(ii)}$$

where $A_0 = \frac{1}{T}\int_0^T y\, dt, A_r = \frac{2}{T}\int_0^T y \cos r\,\omega t\, dt$

and $B_r = \frac{2}{T}\int_0^T y \sin r\,\omega t\, dt.$

Let us obtain the values of the coefficients A_0, A_r and B_r of the Fourier series :

$$A_0 = \frac{1}{T}\int_0^T y\, dt = \frac{a}{T}\int_0^T \left(1 - \frac{t}{T}\right) dt$$

$$= \frac{a}{T}\left[t - \frac{t^2}{2T}\right]_0^T = \frac{a}{T}\left[T - \frac{T^2}{2T}\right] = \frac{a}{2}.$$

which gives the ordinate of the axis of the curve.

For A_r, we have

$$A_r = \frac{2}{T}\int_0^T y \cos r\omega t\, dt$$

$$= \frac{2}{T}\int_0^T y \cos\frac{2r\pi t}{T} dt. \qquad \left[\because \omega = \frac{2\pi}{T}\right]$$

Substituting the value of y from eq. (i), we have

$$A_r = \frac{2}{T}\int_0^T a\left(1-\frac{t}{T}\right)\cos\frac{2r\pi t}{T} dr$$

$$= \frac{2a}{T}\int_0^T \cos\frac{2r\pi t}{T} dt - \frac{2a}{T^2}\int_0^T t\cos\frac{2r\pi t}{T} dt$$

$$= -\frac{2a}{T^2}\int_0^T t\cos\frac{2r\pi t}{T} dt \qquad \left[\because \int_0^T \cos\frac{2r\pi t}{T} dt = 0\right]$$

$$= -\frac{2a}{T^2}\left[t\frac{\sin(2r\pi t/T)}{2r\pi/T} - \int\frac{\sin(2r\pi t/T)}{2r\pi/T}\right]_0^T \quad \text{(by parts)}$$

$$= -\frac{2a}{T^2}\left[t\frac{\sin(2r\pi t/T)}{2r\pi/T} + \frac{\cos(2r\pi t/T)}{(2r\pi/T)^2}\right]_0^T$$

$$= -\frac{2a}{T^2}\left[T\frac{\sin 2r\pi}{2r\pi/T} + \frac{\cos 2r\pi}{(2r\pi/T)^2} - 0 - \frac{\cos 0}{(2r\pi/T)^2}\right]$$

$$= -\frac{2a}{T^2}\left[\frac{1}{(2r\pi/T)^2} - \frac{1}{(2r\pi/T)^2}\right]$$

$$= 0. \qquad [\because \sin 2r\pi = 0 \text{ and } \cos 2r\pi = 1]$$

For B_r, we have

$$B_r = \frac{2}{T}\int_0^T y \sin r\omega t\, dt$$

$$= \frac{2}{T}\int_0^T y \sin\frac{2r\pi t}{T} dt. \qquad \left[\because \omega = \frac{2\pi}{T}\right]$$

Substituting the value of y from eq. (i), we get

$$B_r = \frac{2}{T}\int_0^T a\left(1-\frac{t}{T}\right)\sin\frac{2r\pi t}{T}dt$$

$$= \frac{2a}{T}\int_0^T \sin\frac{2r\pi t}{T}dt - \frac{2a}{T^2}\int_0^T t\sin\frac{2r\pi t}{T}dt$$

$$= -\frac{2a}{T^2}\int_0^T t\sin\frac{2r\pi t}{T}dt \quad \left[\because \int_0^T \sin\frac{2r\pi t}{T}dt = 0\right]$$

$$= -\frac{2a}{T^2}\left[t\cdot -\frac{\cos(2r\pi t/T)}{2r\pi/T} - \int\frac{-\cos(2r\pi t/T)}{2r\pi/T}\right]_0^T \quad \text{(by parts)}$$

$$= -\frac{2a}{T^2}\left[t\frac{\cos(2r\pi t/T)}{2r\pi/T} - \frac{\sin(2r\pi t/T)}{(2r\pi/T)^2}\right]_0^T$$

$$= \frac{2a}{T^2}\left[t\frac{\cos 2r\pi}{2r\pi/T} - \frac{\sin 2r\pi}{(2r\pi/T)^2} - 0 + \frac{\sin 0}{(2r\pi/T)^2}\right]$$

$$= \frac{2a}{T^2}\left[T\frac{1}{2r\pi/T}\right] \quad [\because \cos 2r\pi = 1 \text{ and } \sin 2r\pi = 0]$$

$$= \frac{a}{r\pi}.$$

Thus $B_1 = \frac{a}{\pi}, B_2 = \frac{a}{2\pi}, B_3 = \frac{a}{3\pi}, \ldots$

Substituting the values of A_0, A_r and B_r in eq. (ii) we have

$$y = \frac{a}{2} + \frac{a}{\pi}\left(\sin\omega t + \frac{1}{2}\sin 2\omega t + \frac{1}{3}\sin 3\omega t + \ldots\right)$$

or $$y = \frac{a}{2} + \frac{a}{\pi}\left(\sin\frac{2\pi t}{T} + \frac{1}{2}\sin\frac{4\pi t}{T} + \frac{1}{3}\sin\frac{6\pi t}{T} + \ldots\right).$$

Thus the given periodic motion has the axis $y = \frac{a}{2}$ above the time-axis, and the various components as under

$$\frac{a}{\pi}\sin\frac{2\pi t}{T},$$

$$\frac{a}{2\pi}\sin\frac{4\pi t}{T},$$

$$\frac{a}{3\pi}\sin\frac{2\pi t}{T},$$

and so on, having frequencies in the ratio 1 : 2 : 3 ... and amplitudes in the ratio $1:\frac{1}{2}:\frac{1}{3}:...$

The addition of successive terms of this series is indicated graphically in Fig. It is seen that greater the number of terms used, the closer the resemblances between the resultant curve and the curve under analysis.

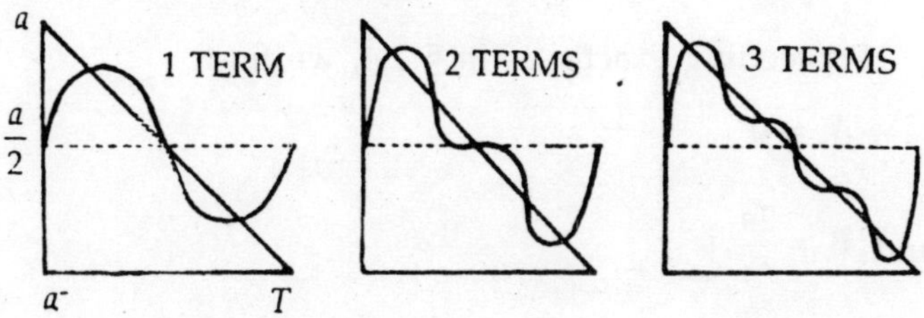

Fig.

Analysis of Saw-Tooth Function

The given function

$$y = a\left(1-\frac{2t}{T}\right), \text{ for } 0 < t < T \qquad \ldots\text{(i)}$$

represents a saw-tooth wave shown in Fig. It can be written as a Fouriers series

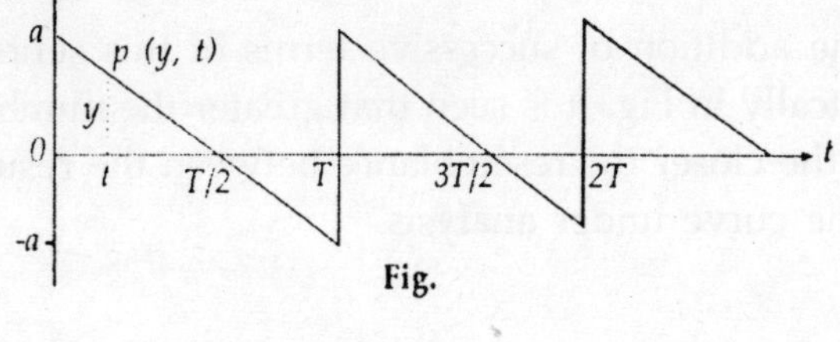

Fig.

$$y = A_0 + A_1 \cos \omega t + ... + A_r \cos r\omega t + ...$$
$$+ B_1 \sin \omega t + ... + B_r \sin r\omega t + ... \qquad \text{(ii)}$$

Let us evaluate the Fourier's coefficients A_0, A_r and B_r.

$$A_0 = \frac{1}{T}\int_0^T y \, dt$$

$$= \frac{1}{T}\int_0^T a\left(1 - \frac{2t}{T}\right)dt$$

$$= \frac{a}{T}\left[t - \frac{t^2}{T}\right]_0^T$$

$$= 0.$$

Proceeding exactly as in § 2.4, we find

$$A_r = 0$$

and $B_r = \dfrac{2a}{r\pi}$.

Thus $B_1 = \dfrac{2a}{\pi}, B_2 = \dfrac{2a}{2\pi}, B_3 = \dfrac{2a}{3\pi}, ...$

Substituting these values in eq. (ii), we get

$$y = \frac{2a}{\pi}\left(\sin \omega t + \frac{1}{2}\sin 2\omega + \frac{1}{3}\sin 3\omega t + ...\right)$$

or $$y = \frac{2a}{\pi}\left(\sin\frac{2\pi t}{T} + \frac{1}{2}\sin\frac{4\pi t}{T} + \frac{1}{3}\sin\frac{6\pi t}{T} + ...\right).$$

The addition of successive terms of this series is shown graphically in Fig. It is seen that greater the number of terms used, the closer the resemblance between the resultant curve and the curve under analysis.

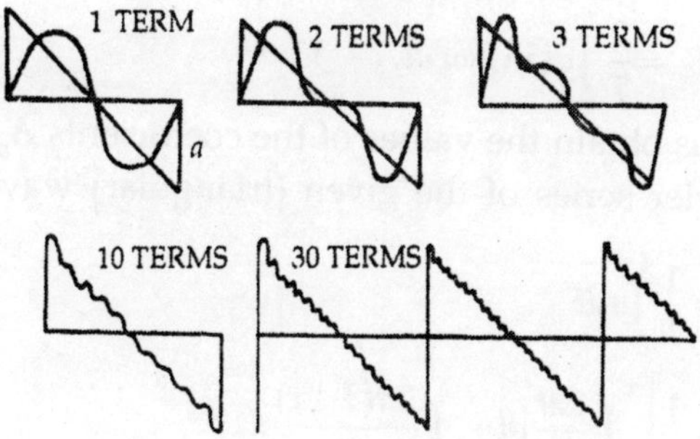

Fig.

(c) *Analysis of A Triangular Wave*

Let us consider triangular wave-form (Fig.),

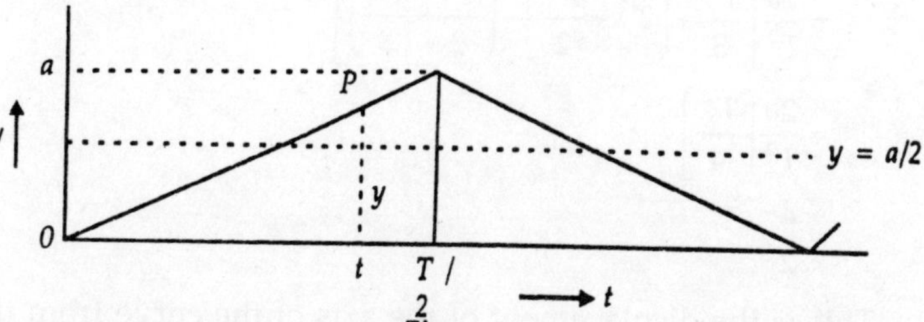

Fig.

and the curve repeats itself after $t = T$. The displacement at any time t may be written as

$$y = \frac{2at}{T} \quad \text{from } t = 0 \text{ to } t = T/2$$

and $$y = \frac{2a(T-t)}{T} \quad \text{from } t = T/2 \text{ to } t = T.$$

Now, the Fourier's series is

$$y = A_0 + A_1 \cos\omega t + ... + A_r \cos r\omega t + ...$$
$$+ B_1 \sin\omega t + ... + B_r \sin r\omega t + ...$$

where $A_0 = \frac{1}{T}\int_0^T y\, dt, A_r = \frac{2}{T}\int_0^T y \cos r\,\omega t\, dt$

and $B_r = \frac{2}{T}\int_0^T y \sin r\omega t\, dt.$

Let us obtain the values of the coefficients A_0, A_r and B_r of the Fourier series of the given (triangular) wave-form :

$$A_0 = \frac{1}{T}\int_0^T y\, dt$$

$$= \frac{1}{T}\left[\int_0^{T/2} \frac{2at}{T} dt + \int_{T/2}^{T} \frac{2a(T-t)}{T} dt\right]$$

$$= \frac{1}{T}\left[\frac{2a}{T}\left(\frac{t^2}{2}\right)_0^{T/2} + \frac{2a}{T}\left(Tt - \frac{t^2}{2}\right)_{T/2}^{T}\right]$$

$$= \frac{2a}{T^2}\left[\frac{T^2}{8} + \left(T^2 - \frac{T^2}{2}\right) - \left(\frac{T^2}{2} - \frac{T^2}{8}\right)\right]$$

$$= \frac{2a}{T^2}\left[\frac{T^2}{4}\right]$$

$$= \frac{a}{2}.$$

This is the displacement of the axis of the curve from the axis of the coordinate system. Now,

$$A_r = \frac{2}{T}\int_0^T y \cos r\omega t\, dt$$

$$= \frac{2}{T}\left[\int_0^{T/2} \frac{2at}{T} \cos r\omega t\, dt + \int_{T/2}^{T} \frac{2a(T-t)}{T} \cos r\omega t\, dt\right]$$

$$= \frac{4a}{T^2}\left[\int_0^{T/2} t \cos r\omega t\, dt + \int_{T/2}^{T} T \cos r\omega t\, dt - \int_{T/2}^{T} t \cos r\omega t\, dt\right]$$

On solving these integrals, we obtain

$$A_r = \frac{4a}{T^2}\left[\frac{2}{r^2\omega^2}\{(\cos \pi r) - 1\}\right]$$

$$= \frac{4a}{T^2}\left[\frac{2}{r^2 (2\pi/T)^2}\{(-1)^r - 1\}\right] \quad [\because \cos \pi r = (-1)^r]$$

$$= \frac{2a}{\pi^2 r^2}\left[(-1)^r - 1\right]$$

$$= 0, \text{ if } r \text{ is even}$$

$$= -\frac{4a}{\pi^2 r^2}, \text{if } r \text{ is odd.}$$

Thus, only odd harmonics with cosine terms would appear in the Fourier series.

$$B_r = \frac{2}{T}\int_0^T y \sin r\,\omega t\, dt$$

$$= \frac{2}{T}\left[\int_0^{T/2} \frac{2at}{T} \sin r\omega t\ dt + \int_{T/2}^{T} \frac{2a(T-t)}{T} \sin r\omega t\, dt\right]$$

$$= \frac{4a}{T^2}\left[\int_0^{T/2} t \sin r\omega t\ dt + \int_{T/2}^{T} T \sin r\omega t\, dt - \int_{T/2}^{T} \sin r\omega t\, dt\right]$$

$$= 0.$$

Thus all the sine terms in the Fourier series are zero.

Substituting the values of A_0, A_r and B_r in the Fourier series, we obtain

$$y = \frac{a}{2} - \frac{4a}{\pi^2}\cos\omega t - \frac{4a}{3^2\pi^2}\cos 3\omega t - \frac{4a}{5^2\pi^2}\cos 5\omega t ...$$

$$= \frac{a}{2} - \frac{4a}{\pi^2}\left(\cos\omega t + \frac{1}{3^2}\cos 3\omega t + \frac{1}{5^2}\cos 5\omega t + ...\right).$$

Thus the given triangular vibration has the axis $y = \frac{a}{2}$ above the time-axis, and the various cosine components of amplitude coefficients,

$$-\frac{4a}{\pi^2}\left(1, \frac{1}{3^2}, \frac{1}{5^2}, \frac{1}{7^2}, ...\right)$$

and frequencies in the ratio 1 : 3 : 5 : 7 ....

The resultant obtained by adding three terms of this series is shown in Fig. As the series is rapidly convergent, the curve under analysis is closely represented by the sum of only a few terms of the series.

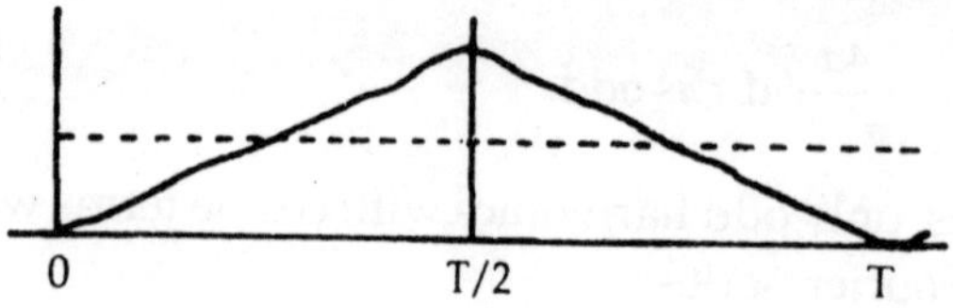

Fig.

(d) *Fourier's Analysis of the Output Wave From A Half-Wave Rectifier*

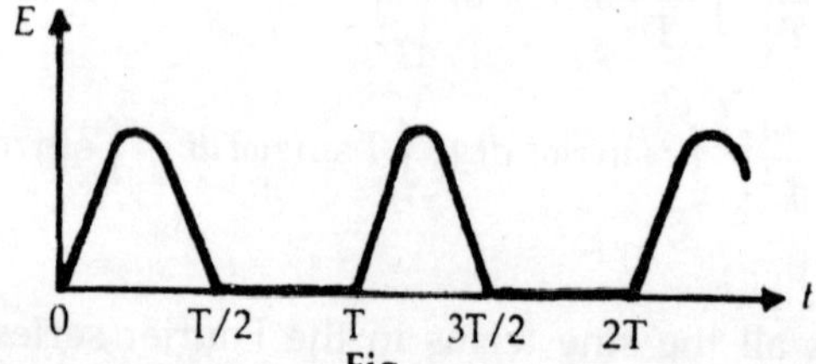

Fig.

A sinusoidal voltage $E = E_0 \sin \omega t$ is passed through a half-wave rectifier which removes the negative half-cycles of the wave. The output voltage wave is of the form shown in Fig. This may be expressed as

$$E(t) = \begin{cases} E_0 \sin \omega t & \text{from } t = 0 \text{ to } t = T/2 \\ 0 & \text{from } t = T/2 \text{ to } t/T \end{cases} \quad T = \frac{2\pi}{\omega}. \qquad \text{...(i)}$$

Let us express it as a Fourier series :

$$E(t) = A_0 + A_1 \cos \omega t + ... + A_r \cos r\omega t + ...$$
$$+ B_1 \sin \omega t + ... + B_r \sin r\omega t + ... \qquad \text{...(ii)}$$

Let us evaluate the Fourier coefficients :

$$A_0 = \frac{1}{T} \int_0^T E(t)\, dt$$

$$= \frac{1}{T} \int_0^{T/2} E_0 \sin \omega t\, dt$$

$$= \frac{1}{T}\int_0^{T/2} E_0 \sin\frac{2\pi t}{T} dt$$

$$= -\frac{E_0}{T}\frac{1}{2\pi/T}\left[\cos\frac{2\pi t}{T}\right]_0^{T/2}$$

$$= -\frac{E_0}{2\pi}[\cos\pi - \cos 0]$$

$$= \frac{E_0}{\pi}. \quad [\because \cos\pi = -1 \text{ and } \cos 0 = 1]$$

Again,

$$A_r = \frac{2}{T}\int_0^T E(t)\cos r\omega t\, dt$$

$$= \frac{2E_0}{T}\int_0^{T/2} \sin\omega t \cos r\omega t\, dt$$

$$= \frac{2E_0}{T}\int_0^{T/2} \sin\frac{2\pi t}{T}\cos\frac{2r\pi t}{T} dt$$

$$= \frac{2E_0}{T}\int_0^{T/2} \frac{1}{2}\left[\sin(1+r)\frac{2\pi t}{T} + \sin(1-r)\frac{2\pi t}{T}\right] dt$$

$$= -\frac{E_0}{T}\left[-\frac{\cos(1+r)2\pi t/T}{(1+r)2\pi/T} - \frac{\cos(1-r)2\pi t/T}{(1-r)2\pi/T}\right]_0^{T/2}$$

$$= -\frac{E_0}{T}\left[\frac{\cos(1+r)\pi - \cos 0}{(1+r)2\pi/T} + \frac{\cos(1-r)\pi - \cos 0}{(1-r)2\pi/T}\right]$$

$$= -\frac{E_0}{2\pi}\left[\frac{\cos(1+r)\pi - 1}{(1+r)} + \frac{\cos(1-r)\pi - 1}{(1-r)}\right].$$

When r is odd then this is equal to zero, because $\cos 0 = \cos 2\pi = \cos 4\pi = 1$. When r is even, then we have

$$A = \frac{E_0}{2\pi}\left[\frac{2}{1+r} + \frac{2}{1-r}\right] \quad [\because \cos\pi = \cos 3\pi = \cos 5\pi = -1]$$

$$= \frac{E_0}{\pi}\frac{2}{(1+r)(1-r)}$$

$$= -\frac{2E_0}{\pi}\frac{1}{(r-1)(r+1)};\ r = 2,4,6,...$$

$$= -\frac{2E_0}{\pi}\left[\frac{1}{(1)(3)}, \frac{1}{(3)(5)}, \frac{1}{(5)(7)}, ...\right].$$

Again,

$$B_r = \frac{2}{T}\int_0^T E(t)\sin r\omega t\, dt$$

$$= \frac{2E_0}{T}\int_0^{T/2} \sin\frac{2\pi t}{T}\sin\frac{2r\pi t}{T} dt$$

Proceeding as above, we can see that

$$B_r = \frac{E_0}{2} \text{ for } r = 1$$

and $B_r = 0$ for $r = 2,3,4,....$

Substituting these values of A_0, A_r, and B_r in eq. (ii), we get

$$E(t) = \frac{E_0}{\pi} + \frac{E_0}{2}\sin\omega t - \frac{2E_0}{\pi}\left[\frac{1}{1.3}\cos 2\omega t + \frac{1}{3.5}\cos 4\omega t + \frac{1}{5.7}\cos 6\omega t + ...\right].$$

(e) *Fourier Analysis of a Full Wave Rectifier Output*

The output voltage (or current) of a full-wave rectifier voltage is continuous, uni-directional but pulsating. It can be analysed by means of Fourier series.

Let $E = E_0 \sin \omega t$ be the applied input voltage to the rectifier. It r_p be the internal plate resistance of the diode, then the instantaneous output current through a load resistance R will be given by

$$i = \frac{E}{r_p + R} = \frac{E_0 \sin\omega t}{r_p + R}$$

$$= i_0 \sin\omega t,$$

where i_0 $\{=E_0\ (r_p + R)\}$ is the peak value of the current in the load.

Taking $f(\omega t) = i + i_0 \sin \omega t$, we can evaluate the coefficients A_0, A_r, B_r.

In the case of full-wave rectifier, i is positive for $t = 0$ to $t = T/2$, as well as for $t = T/2$ to $t = T$. That is,

$$i = i_0 \sin \omega t \quad \text{for} \quad t = 0 \text{ to } t = T/2$$

and $$i = -i_0 \sin \omega t \quad \text{for} \quad t = T/2 \text{ to } t = T,$$

because sin ωt is negative in the range t = T/2 to t = T. Now, we have

$$i = f(t) = A_0 + \sum_{r=1}^{\infty}\left[A_r \cos(\omega t r) + B_r \sin(\omega r t)\right]$$

where

$$A_0 = \frac{2}{T}\left[\int_0^{T/2} i \sin \omega t\, dt - \int_{T/2}^{T} i \sin \omega t\, dt\right]$$

$$= \frac{\omega}{2\pi}\left[\int_0^{\pi/\omega} i_0 \sin \omega t\, dt - \int_0^{2\pi/\omega} i_0 \sin \omega t\, dt\right]$$

$$= \frac{\omega}{2\pi}\frac{i_0}{\omega}\left[\left|-\cos \omega t\right|_0^{\pi/\omega} - \left|\cos \omega t\right|_{\pi/\omega}^{2\pi/\omega}\right]$$

$$= \frac{i_0}{2\pi}\left[(-\cos \pi + \cos 0) - (-\cos 2\pi + \cos \pi)\right]$$

$$= \frac{2i_0}{2\pi}\left[(1+1) - (-1-1)\right]$$

$$= \frac{2i_0}{\pi}$$

$$A_r = \frac{2}{T}\left[\int_0^{T/2} i_0 \sin \omega t \cos r\omega t\, dt - \int_0^{T} i_0 \sin \omega t \cos r\omega t\, dt\right]$$

$$= \frac{\omega}{\pi}\frac{i_0}{2}\left[\int_0^{\pi/\omega} \{\sin(r+1)\omega t - \sin(r-1)\omega t\}\, dt\right.$$

$$\left. - \int_{\pi/\omega}^{2\pi/\omega} \{\sin(r+1)\omega t - \sin(r-1)\omega t\}\, dt\right]$$

$$= \frac{\omega}{\pi}\frac{i_0}{2\omega}\left[\left| -\frac{\cos(r+1)\omega t}{(r+1)\omega} + \frac{\cos(r-1)\omega t}{(r-1)\omega}\right|_0^{\pi/\omega}\right.$$

$$\left. - \left| -\frac{\cos(r+1)\omega t}{(r+1)\omega t} + \frac{\cos(r-1)\omega t}{(r-1)\omega}\right|_{\pi/\omega}^{2\pi/\omega}\right]$$

$$= \frac{\omega}{\pi}\frac{i_0}{2\omega}\left[\left\{ -\frac{\cos(r+1)\pi}{r+1} + \frac{\cos(r-1)\pi}{r-1} + \frac{\cos 0}{r+1} - \frac{\cos 0}{r-1}\right\}\right.$$

$$\left. - \left\{ -\frac{\cos(r+1)2\pi}{r+1} + \frac{\cos(r-1)2\pi}{r-1} + \frac{\cos(r+1)\pi}{r+1} - \frac{\cos(r-1)\pi}{r-1}\right\}\right]$$

Now, when r is odd

$$A_r = \frac{i_0}{2\pi}\left[\left\{ -\frac{1}{r+1} + \frac{1}{r-1} + \frac{1}{r+1} - \frac{1}{r-1}\right\}\right.$$

$$\left. - \left\{ -\frac{1}{r+1} + \frac{1}{r-1} + \frac{1}{r+1} - \frac{1}{r-1}\right\}\right]$$

$$= 0$$

When r is even :

$$A_r = \frac{i_0}{2\pi}\left[\left\{ -\frac{1}{r+1} - \frac{1}{r-1} + \frac{1}{r+1} - \frac{1}{r-1}\right\}\right.$$

$$\left. - \left\{ -\frac{1}{r+1} + \frac{1}{r-1} - \frac{1}{r+1} + \frac{1}{r-1}\right\}\right]$$

$$= \frac{4i_0}{2\pi}\left[\frac{1}{r+1} - \frac{1}{r-1}\right]$$

$$= -\frac{4i_0}{\pi(r^2-1)}$$

Thus the values of the coefficients A's are (r = 2, 4, 6, ...)

$$-\frac{4i_0}{3\pi}, -\frac{4i_0}{15\pi}, -\frac{4i_0}{35\pi}, \ldots$$

$$B_r = \frac{2}{T}\left[\int_0^{T/2} i_0 \sin\omega t \sin r\omega t\, dt - \int_{T/2}^{T} i_0 \sin\omega t \sin r\omega t\, dt\right]$$

$= 0$ (it can be worked out exactly as above).

Thus the output current i can be expressed in the form of a Fourier series as

$$i = \frac{2i_0}{\pi} - \frac{4i_0}{3\pi}\cos 2\omega t - \frac{4i_0}{15\pi}\cos 4\omega t - \frac{4i_0}{35\pi}\cos 6\omega t ...$$

Thus it consists of a d.c. component $\frac{2i_0}{\pi}$ and a series of a.c. components (even harmonics only) or ripples.

Summing of Infinite Series

(a) ***Prove*** $\frac{\pi^2}{6} = 1 + \frac{1}{2^2} + \frac{1}{3^2} + \frac{1}{4^2} + ...$ ...(i)

Let $f(x) = x^2, -\pi < x < \pi$

This is an even function and therefore all the sine terms in its Fourier expansion are zero

So $f(x) = a_0 + \sum_{n=1}^{\infty} a_n \cos nx,$

where $a_0 = \frac{1}{2\pi}\int_{-\pi}^{\pi} x^2 dx = \frac{\pi^2}{3}$ ...(ii)

$$a_n = \frac{2}{\pi}\int_0^{\pi} x^2 \cos n\, x\, dx = (-1)^n \frac{4}{n^2} \quad ...(iii)$$

Thus $x^2 = \frac{\pi^2}{3} + 4\left[-\cos x + \frac{\cos 2x}{4} - \frac{\cos 3x}{9} + ...\right]$ ...(iv)

For $x = \pi$, this reduces to

$$x^2 = \frac{\pi^2}{3} + 4\left[1 + \frac{1}{4} + \frac{1}{9} + ...\right]$$

or $\frac{\pi^2}{6} = \frac{1}{1} + \frac{1}{2^2} + \frac{1}{3^2} + ...$ ...(v)

Hence the result.

(b) *Determine the Fourier series expansion of the Function*

$$f(x) = \pi^2 - x^2 \quad \text{for} \quad -\pi \le x \le \pi$$

Here the time period is 2π.

The Fourier series will be given as

$$\frac{A_0}{0} + \sum_{n=1}^{\infty}\left[A_n \cos(nx) + B_n \sin(nx)\right]$$

where the Fourier coefficients A_0 is given by

$$A_0 = \frac{1}{\pi}\int_{-n}^{n}(\pi^2 - x^2)\,dx = \frac{4\pi^2}{3}$$

and
$$A_n = \frac{1}{\pi}\int_{-n}^{n}(\pi^2 - x^2)\cos(nx)\,dx \quad \text{for } n = 1, 2, \ldots$$

$$= \frac{1}{\pi}\left\{\int_{-n}^{0}(\pi^2 - x^2)\cos(nx)\,dx + \int_{0}^{\pi}(\pi^2 - x^2)\cos(nx)\,dx\right\}$$

(Change $x \to -x$)

$$= \frac{1}{\pi}\left\{-\int_{-n}^{0}(\pi^2 - x^2)\cos(-nx)\,dx + \int_{0}^{\pi}(\pi^2 - x^2)\cos(nx)\,dx\right\}$$

$$= \frac{1}{\pi}\left\{\int_{0}^{\pi}(\pi^2 - x^2)\cos(nx)\,dx + \int_{0}^{\pi}(\pi^2 - x^2)\cos(nx)\,dx\right\}$$

$$= \frac{2}{\pi}\int_{0}^{\pi}(\pi^2 - x^2)\cos(nx)\,dx$$

Integrating by parts twice we get

$$A_n = \frac{2}{\pi}(\pi^2 - x^2)\left.\frac{\sin(nx)}{n}\right|_0^{\pi} - \frac{2\pi}{n}\int_0^{\pi}(-2x)\sin(nx)\,dx$$

$$= 0 + \frac{4}{\pi n}\left.\frac{x(-\cos nx)}{n}\right|_0^{\pi} + \frac{4}{\pi n^2}\int_0^{\pi}1.\cos(nx)\,dx$$

$$= -\frac{4\pi}{\pi n^2}\cos(n\pi) + \frac{4}{\pi n^3}\left.\sin(nx)\right|_0^{\pi} + \frac{4}{n^2}(-1)^{n+1} + 0$$

$$\therefore \quad A_n = \frac{4(-1)^{n+1}}{n^2} \quad \text{for} \quad n = 0, 1, 2, \ldots$$

Now

$$B_n = \frac{1}{\pi}\int_{-\pi}^{\pi}(\pi^2 - x^2)\sin(nx)\,dx \quad \text{for} \quad n = 1, 2, \ldots$$

Changing $x \to -x$, we have

$$B_n = \frac{1}{\pi}\int^{-\pi}(\pi^2 - (-x)^2)\sin(-nx)(-dx)$$

$$= +\frac{1}{\pi}\int_{\pi}^{-\pi}(\pi^2 - x^2)\sin(nx)(dx)$$

$$= -\frac{1}{\pi}\int_{-\pi}^{-\pi}(\pi^2 - x^2)\sin(nx)(dx)$$

$$= -B_n$$

so $\quad 2B_n = 0$

or $\quad B_n = 0$

Thus the series is

$$f(x) = \frac{2\pi^2}{3} + 4\left(\cos x - \frac{1}{2^2}\cos 2x + \frac{1}{3^2}\cos 3x + \ldots\right)$$

putting $x = 0$, we have

$$\pi^2 = \frac{2\pi^2}{3} + 4\left(1 - \frac{1}{2^2} + \frac{1}{3^2} - \frac{1}{4^2} + \ldots\right)$$

or

$$\frac{\pi^2}{12} = 1 - \frac{1}{2^2} + \frac{1}{3^2} - \frac{1}{4^2} + \frac{1}{5^2} - \ldots \qquad \ldots(1)$$

Again putting $x = \pm\pi$

$$0 = \frac{2\pi^2}{3} + 4\left(\cos\pi - \frac{1}{2^2}\cos 2\pi + \frac{1}{3^2}\cos 3\pi\ldots\right)$$

or

$$0 = \frac{2\pi^2}{3} - 4\left(1 + \frac{1}{2^2} + \frac{1}{3^2} - \frac{1}{4^2} + \ldots\right)$$

$\therefore$

$$\frac{\pi^2}{6} = 1 + \frac{1}{2^2} + \frac{1}{3^2} + \frac{1}{4^2} + \ldots \qquad \ldots(2)$$

Adding (1) and (2), we have

$$\frac{\pi^2}{4} = 2 + \frac{2}{3^2} + \frac{2}{5^2} + \ldots$$

or
$$\frac{\pi^2}{8} = 1 + \frac{1}{3^2} + \frac{1}{5^2} + \frac{1}{7^2} + \ldots$$

(c) *Determine the Fourier series expansion of the function* $f(x) = |x|$ for $-\pi \le x \le \pi$.

Hence show that $\frac{\pi^2}{8} = 1 + \frac{1}{3^2} + \frac{1}{5^2} + \frac{1}{7^2} + \ldots$

Now $f(-x) = |-x| = |x| = f(x)$ and so is even.

Therefore it can be expanded only in cosine series and the coefficients $A_0, A_1, A_2, \ldots$ can be evaluated from half range only i.e. the Fourier series is

$$f(x) = \frac{A_0}{2} + \sum_{n=1}^{\infty} A_n \cos(nx)$$

where

$$A_0 = \frac{2}{\pi}\int_0^{\pi} |x|\,dx = \frac{2}{\pi}\int_0^{\pi} x\,dx$$

$$= \frac{2}{\pi}\left.\frac{x^2}{2}\right|_0^{\pi} = \frac{\pi^2}{\pi} = \pi$$

$$A_n = \frac{2}{\pi}\int_0^{\pi} |x|\cos(nx)\,dx$$

$$= \frac{2}{\pi}\int_0^{\pi} x\cos(nx)\,dx$$

$$= \frac{2}{\pi}\left.\frac{x\sin(nx)}{(-n)}\right|_0^{\pi} - \frac{2}{\pi n}\int_0^{n} (1)\sin(nx)\,dx$$

$$= -\frac{2}{\pi n}\frac{\cos(nx)}{(-n)}\Bigg|_0^\pi = \frac{2}{n^2}\left[(-1)^n - 1\right]$$

$$= \frac{2}{\pi n^2}\left[(-1)^n - 1\right] \quad \text{for} \quad n = 1, 2, 3, \ldots$$

$$\therefore \quad A_1 = -\frac{4}{1^2\pi},\ A_2 = \frac{1}{2^2}(1-1) = 0$$

$$A_3 = -\frac{4}{3^2\pi},\ A_4 = 0,\ A_5 = -\frac{4}{5^2\pi},\ \ldots$$

$$|x| = \frac{\pi}{2} - \frac{4}{1^2\pi}\cos x - \frac{4}{\pi 3^2}\cos 3x - \frac{4}{5^2\pi}\cos 5x - \ldots$$

for $x = \pm\pi$, we have

$$\pi = \frac{\pi}{2} + \frac{4}{\pi}\left(1 + \frac{1}{3^2} + \frac{1}{5^2} + \frac{1}{7^2} + \ldots\right)$$

$$\therefore \quad \frac{\pi^2}{8} = 1 + \frac{1}{3^2} + \frac{1}{5^2} + \frac{1}{7^2} + \ldots$$

Bibliography

Ablowitz, Mark J. : *Complex Variables : Introducation and Applications*, Routledge, London, 2000.

Alexanderson, Gerald : *The William Lowell Putnam Mathematical Competition : Problems and Solutions*, University of Santa Clara, Santa Clara, 2004.

Anderson, Marlow and Robin Wilson : *Sherlock Holmes in Babylon and Other Tales of Mathematical History*, The Open University, Milton Keynes, 2004.

Angus, Robert B. : *Math at Work*, Prentice Hall, London, 1999.

Anthony, Martin : *Mathematics for Economics and Finance*, Routledge, London, 1995.

Arnold, R, and Others : *Excursions in Modern Mathematics*, Prentice Hall, London, 1999.

Baker, Andrew : *Structured Ring Spectra*, Universty of Glasgow, Glasgow, 2004.

Baker, H. F. : *Multiply Periodic Functions*, Cambridge Mathematical Library, Cambridge, 2004.

Bemier, G. and Kinet J.M. : *Thermal and Statistical Physics*, C.R.C. Press, Boca Raton, 1981.

Berlinghoff, William P. : *Math through the Ages : a Gentle History for Teachers and Others*, Colby College, Maine, 2004.

Bhattacharya, P. B. : *Basic Abstract Algebra*, Cambridge University Press, New York, 1998.

Billingham, J., King, A. C. and S. R. Otto : *Differential Equations*, Blackwell Publishing, London, 1995.

Bollobas, Bila : *Linear Analysis : An Introductory Course*, Routledge, London, 1991.

Bolt, Brian : *A Mathematical Jamboree*, Routledge, London, 1992.

—— : *A Mathematical Pandora's Box*, Blackwell Publishing House, London, 1993.

Bray, C. : *Nitrogen Metabolism in Plants*, Longman, London, 1984.

Brown, B. W. : *Finite Mathematics*, Ardsley Publishing House, London, 1992.

Buhlmann, Hans : *Mathematical Methods in Risk Theory*, Springer, London, 1996.

Cams, H.R. : *Elements of Thermal and Statistical Physics*, Oxford University Press, London, 1966.

Casselman, William : *Mathematical Illustrations*, University of British Columbia, Vancouver, 2004.

Cleaves, Cheryl : *College Mathematics for Technology*, Prentice Hall, London, 1995.

Corfield, David : *Towards a Philosophy of Real Mathematics*, Universty of Oxford, Oxford, 2003.

Cristina, Pedicchio Maria : *Encyclopaedia of Mathematics and its Applications*, York University, Toronto, 2004.

Dershowitz, N. and Others : *Calendrical Calculations*, Cambridge University Press, Cambridge, 1998.

Devenport, H. : *The Higher Arithmetic*, Cambridge, London, 1999.

Duffy, Dean G. : *Advanced Engineering Mathematics*, CRC Press, New York, 1998.

Edixhoven, Bas : *Compositio Mathematica*, Leiden Unversity, Leiden, 2004.

Eichoiz, Robert E. : *Addison-Wesley Mathematics,* Add. Wesley, London, 1989.

Epstein, E. : *Thermal Physics : Principles and Perspectives,* John Wiley, New York, 1972.

Feferman, Burdman Anita and Feferman, Solomon : *Alfred Tarski : Life and Logic,* Stanford University, California, 2004.

Frank, Harry : *Statistics,* Routledge Publishing House, London, 1990.

Green, J.R. : *An Introduction to Thermal and Statistical Physics,* MacMillan, London, 1911.

Gregory, Phil : *Bayesian Logical Data Analysis for the Physical Sciences,* University of British Columbia, Vancouver, 2004.

Hardy, G. H. : *A Mathematician's Apology,* Cambridge University, London, 2001.

Jasmer, Singh : *Solid Geometry,* Dominant Publishers, New Delhi, 2003.

Joshi, Mark S. : *Mathematic Finance and Risk,* Royal Bank of Scotland, Scotland, 2003.

—— : *The Concepts and Practice of Mathematical Finance,* Royal Bank of Scotland, Scotland, 2003.

Keddy, Bittinger and Beecher : *Developmental Mathematics,* D' Books, Harcourt Brace, 1993.

Kirkby, E.A. : *Mathematical Physics,* The University of Leeds, Leeds, 1970.

Kumar, Sunil : *Abstract Algebra,* Campus Books, New Delhi, 2004.

—— : *Advanced Calculus,* Campus Books, New Delhi, 2004.

—— : *Modern Algebra,* Campus Books, New Delhi, 2005.

Lial, M. L . and Others : *Beginning Algebra,* Add. Wesley, London, 2000.

—— : *Basic College Mathematics,* Add. Wesley, London, 1998.

Lindsey, J. K. : *Statistical Analysis of Stochastic Processes in Time,* Universite de liege, Belgium, 2004.

Lyons, Louis : *Mathematics for Science Students,* Cambridge University Press, New York, 1991.

Manton, Nicholas : *Topological Solitons,* University of Kent, Kent, 2004.

Marchioro, Carlo and Mario : *Mathematical Theory of Incompressible Nonviscous Fluids,* Springer, London, 1994.

Mason, A. W. : *Glasgow Mathematical,* Universty of Galsgow, Galsgow, 2004.

McCauley, Joseph L. : *Dyanamics of Markets,* University of Houston, Houston, 2004.

Mora, Teo : *Encyclopeadia of Mathematics and its Applications,* University of Genoa, Genoa, 2003.

Motwani, Rajiv and Raghavan, Prabhakar : *Randomized Algorithms,* Cambridge University Press, New York, 1994.

Muller-Stach, S. : *Transcendental Aspects of Algebraic Cycles,* Cambridge University Press, Cambridge, 2001.

Nesetril, Jaroslav : and Rodi, Vojtech : *Mathematics of Ramsey Theory,* Springer, London, 1990.

Ovsienko, V. : *Projective Differential Geometry Old and New,* Pennsylvania State University, Pennsylvania, 2004.

Pande, Srivastava and Gupta : *A Textbook of Algebra and Trigonometry,* Dominant Publishers, New Delhi, 2000.

Parker, Leonard, and Others : *Math Tensor : A System for Doing Tensor Analysis by Computer,* Add. Wesley, London, 1994.

Parks, Harold and Others : *Mathematics in Life Society and the World,* Prentice Hall, London, 1997.

Rabinowitch, E. and Govindjee : *Thermodynamics and Statistical Physics,* John Wiley, New York, 1969.

Ramshaw, J.A.M. : *Thermodynamics Physics,* Oxford University Press, New York, 1982.

Rawat, K. S. : *A Textbook of Calculus with Differential Equation*, Campus Books, New Delhi, 2001.

Ray, Subhash C. : *Data Envelopment Analysis*, University of Connecticut, Connecticut, 2004.

Riley, K. F. : *Mathematical Methods for Physics and Engineering*, Blackwell Publishing House, London, 1990.

Ross, Sheldon M. : *An Elementary Introduction to Mathematical Finance*, Cambridge University Press, New York, 1995.

Russel, John : *Dictionary of Mathematics*, Campus Books, New Delhi, 2004.

—— : *Teaching of Mathematics*, Campus Books, New Delhi, 2004.

Salisbury, F.B. and Ross, C.W. : *Thermal Physics*, Woodworth Publishing Co., California, 1967.

Schechter, Martin : *An Introduction to Nonlinear Analysis*, University of California, Irvine, 1995.

Sharma, A. K. : *Encyclopaedia of Analytical Geometry*, Dominant Publishers, New Delhi, 2001.

—— : *Encyclopaedia of Mathematical Analysis*, Dominant Publishers, New Delhi, 2001.

Singh, Srivastava and Siddiqui : *A Textbook of Calculus*, Dominant Publishers, New Delhi, 2003.

—— : *Intergral Calculus*, Dominant Publishers, New Delhi, 2000.

Singh and Denis : *Special Functions and Their Applications*, Dominant Publishers, New Delhi, 2001.

Snieder, Roel : *A Guided Tour of Mathematical Methods*, Colorado School of Mines, Colorado, 2004.

Strien, Van S. : *Ergodic Theory and Dynamical Systems*, University of Warwick, Warwick, 2004.

Szekeres, Peter : *A Course in Modern Mathematical Physcis*, University of Adelaide, Adelaide, 2004.

Taylor, Alan D. : *Social Choice and the Mathematics of Manipulation*, Union College, New York, 2004.

Thomas, C. B. : *Mathematical Proceedings of the Cambridge Philosophical Society*, Universty of Cambridge, Cambridge, 2004.

Tikhonov, A. N. and Others : *Differential Equations*, Springer Publishers, New York, 1985.

Trivedi, S. D. : *Comprehensive Differential Equations*, Dominant Publishers, New Delhi, 2003.

Uppreti, D. : *Advanced Dictionary of Mathematics*, Dominant Publishers, New Delhi, 1999.

Van, Lint J.H. : *A Course in Combinatorics*, Blackwell Publishing, House, London, 1990.

Verner, J.E. : *Thermal Physics*, Academic Press, New York, 1985.

Waldo, Dunnington G. : *Gauss : Titan of Science*, The Open University, Milton Keynes, 2004.

William, M. Setek and Others : *Fundamentals of Mathematics*, Prentice Hall, London, 1999.

Wilson, Robin J. : *Topics in Algebraic Graph Theory*, The Open University, Milton Keynes, 2004.

Wisniewski, M. : *Introductory Mathematical Methods in Economics*, McGraw Hill, London, 1996.

Zabezyk, Jerzy : *Mathematical Control Theory : An Introduction*, Birkhauser, 1998.

Zeevart, J.A. : *Mathematical Physics*, McGraw Hill, London, 1976.